Microbial Aspects of Metallurgy

Microbial Aspects of Metallurgy

EDITOR: J. D. A. MILLER

Lecturer, Corrosion Group, Department of Chemical Engineering, University of Manchester Institute of Science and Technology

MTP

Medical and Technical Publishing Co. Ltd.
CHILTERN HOUSE AYLESBURY ENGLAND

ISBN-13: 978-94-011-6117-6 e-ISBN-13: 978-94-011-6115-2
DOI: 10.1007/978-94-011-6115-2

Softcover reprint of the hardcover 1st edition 1971

Published in Great Britain by MTP (Medical and Technical Publishing Co. Ltd.) Chiltern House, Aylesbury, 1971
Made and Printed in Great Britain by
Eyre & Spottiswoode Limited, Thanet Press, Margate

SBN 85200 009 X

Contents

CHAPTER ONE: J. D. A. MILLER

Introductory Microbiology

Towards a Definition of a Microbe

For the purposes of this book we shall need to know what microbes are, what they do (which means in effect what chemical reactions go on inside them, and hence their chemical interaction with their environment), and how their growth can be both encouraged and discouraged – mainly the latter. These aspects will be covered in the present chapter, ending with a brief résumé of their role in metallic corrosion and associated activities.

What Characterizes a Living Organism?

Microbes – often called *micro-organisms* or sometimes merely *organisms* when from the context it is clear that small organisms are being referred to – are, unfortunately, not especially easy to define. In attempting to do so, we must remember that the practice of classifying objects – of putting them as it were into labelled pigeon-holes – appears to be fundamental to human beings; and though this general process of reducing chaos to some semblance of order obviously has much to recommend it, the names which some of the biological pigeon-holes bear were coined long before we had any real body of scientific knowledge of living things, or any tools for acquiring such knowledge. They thus need have no validity.

First, then, we should decide which living organisms are microbes and hence are the province of microbiologists, while avoiding if possible the time-honoured "animal" and "plant" pigeon-holes. And to do this we should be clear what a "living organism" is – a point which at first sight may seem too obvious to warrant discussion, but which in fact serves as a good illustration of a general biological principle: that every criterion we devise for defining or classifying things appears to break down at some point.

The vast majority of chemical changes that occur on the earth's surface – both on land and in water – are brought about by the activities of living organisms; and metallic corrosion, to which the larger part of this volume is devoted, has its share of biological causes. The kinds of process by which microbes accomplish these environmental changes are to be outlined later in this chapter; but overall they amount to local reversals of the trend defined by the second law of thermodynamics – an increased state of orderliness and complexity within the living organism over that in the environment, with which must inevitably go a compensating degradation or increase in entropy of some of the environmental material. The ability to carry out chemical reactions of this type is an essential and unique property of living organisms; and these processes must be co-ordinated and directed ultimately towards the growth and reproduction of the organism.

Then what about dormant seeds and spores, freeze-dried bacteria – and probably in the near future, deep-frozen investors awaiting a gigantic return on their holdings? These show no chemical activity: they are in a state of suspended animation, and are really only "living" in the sense that with suitable treatment, or the onset of suitable conditions, they will start, or restart, chemical activities typical of living things. They are merely *viable*.

Now all living things are composed of *cells*, the simplest organisms being unicellular. Some of the unicellular organisms are, at least to all appearances under the microscope, very simple and devoid of structures (organelles); though of course the cell that comprises the organism is highly organized in the chemical and physiological sense, being able to perform every process typical of that organism at the appropriate time in its life-cycle and in response to the appropriate external conditions.

In a very large, complex organism, on the other hand, there is a greater or lesser degree of differentiation of function of the cells comprising it, no one cell being able to emulate the functions of the whole organism. Hence we find that the epithet "living" means different things when applied to a single-cell organism and to a large multicellular organism. In the extreme case, a large animal, the overall organization is such that the "death", *i.e.*, cessation of function for some reason, of let us say the heart or the brain, results almost instantaneously in the death of that creature as an organized, functional unit; though as we know, suitable stimulation of the heart can result in the animal's resuming normal life in certain cases, thus proving that all or most of its individual cells had remained viable for a short period after co-ordination ceased. Similarly, individual tissues (composed of cells of specialized function) can be perpetuated as a tissue culture in a suitable growth medium if taken immediately from a "dead" animal.

To summarize: living organisms are highly organized at least up to the cellular level; and, by assimilating material from their environment they are able to add to their own cell material and eventually reproduce their kind. Viruses are strictly not living things but appear to be aberrant *components* of cells; enzymes (see below) are not living, but are chemical substances with catalytic properties and are found in living cells.

Microbes and Protists

The higher forms of life have of course been known to man since the dawn of human history, and it is not difficult to imagine that the distinction between "animal" and "plant", as applied to these organisms, soon established itself in man's mind. The differences in these cases are clear-cut: a higher plant obtains its nutriment by diffusion of solutes from the soil and gases from the air so that the ability to move would bestow relatively little advantage; animals take in solid food, to find which they must be motile with a few exceptions. Higher plants are green, at least in parts, owing to the presence of chlorophyll, and they have no particular adult size or shape except within rather broad limits; higher animals are rarely green (and those that are do not photosynthesize), they have a fairly definite adult size and shape, the various parts of the body bearing a closely-defined weight (or volume) ratio to one another, and they have many more specialized types of tissue. These two types of life constitute, as it were, well-separated twin apices of an evolutionary tree, and it is quite irrational to suppose that on tracing these organisms back through simpler forms we should find the same clear-cut differences between plants and animals, or indeed, that our invented terms "plant" and "animal" should have any real meaning when extended into the realms of the smaller, structurally simpler organisms.

Our first attempt at a definition of a micro-organism, then, might be a creature that has not necessarily all of the characteristics of an "animal" or a "plant" as described above (it may show some characteristics of both groups, as exemplified by the motility of certain chlorophyll-containing algae and algal spores, the ability of the chlorophyllous unicell *Euglena* to ingest solid particles of food and the inability of colourless forms such as bacteria, yeasts and moulds to do this), and is visible only under the microscope. Since, however, the cells in our bodies are only visible under the microscope, we are at once prompted to stipulate that the cells of the micro-organism should either be separate one from another, or, if joined in some way, they should still be autonomous so that the structure they comprise should have no properties that the individual cells lack: there should be no control superimposed on them from another part of the organism.

Several objections to this definition immediately arise. Unicellular organisms (the true bacteria and certain algae, for instance) are clearly microbes; filamentous algae should therefore be included; but what about the red marine algae which form large tufts, and then the brown marine algae which appear to have a definite structure consisting of holdfast, stipe and frond – and to take the extreme case, the giant kelps of the Pacific, which may reach upwards of 150 ft in length? Where do we put the dividing line between micro-organisms and the rest?

Again, the moulds are simple fungi that grow as colonies of tangled mycelium, composed in turn of undifferentiated, threadlike hyphae analogous to the cells of other microbes (see further on). Small portions of such hyphae, if removed, will themselves develop into colonies of mould on a suitable substratum. The "higher fungi", however, though similarly consisting of a loosely tangled mass of hyphae, will under certain conditions produce a large spore-bearing organ (a mushroom or toadstool), easily visible to the naked eye and obviously having an appreciable degree of supracellular organization. Nevertheless, large though it is, this spore-bearing body is composed merely of compressed masses of hyphae, not of true tissue such as characterizes the higher animals and plants.

Considerations such as those above show that the term "microbe" has only limited usefulness; and this led the German biologist Haeckel as long ago as 1894 to propose, in addition to animals and plants, the creation of a third kingdom: the Protista or protists. This third kingdom contains, besides truly microscopic forms of life, many multicellular organisms including the very large marine forms already mentioned which, owing to their prostrate habit in a solution containing all the nutrients they require, have no need for the differentiation of tissues essential to all but the smallest terrestrial organisms to give them strength and rigidity.

The Protista thus form a relatively natural group which consists of the bacteria, fungi (including yeasts), algae and Protozoa. It can also be divided into those organisms (bacteria and blue-green algae – all microscopic) which have no clearly defined nucleus within the cell, and those which have one or more nuclei per cell, each clearly demarcated during most of the life cycle of the cell by a nuclear membrane. These two groups are called respectively the lower and higher protists.

Having touched on some of the more real basic divisions of the world of living things, and discovered that the term "microbe" has no exact meaning (comprising the lower protists, and more or fewer of the higher protists according to personal taste; though all the organisms that concern us in this book could by common consent be called microbes), we will now examine some of the principal types of organisms in this loose group. Fig. 1.1, which

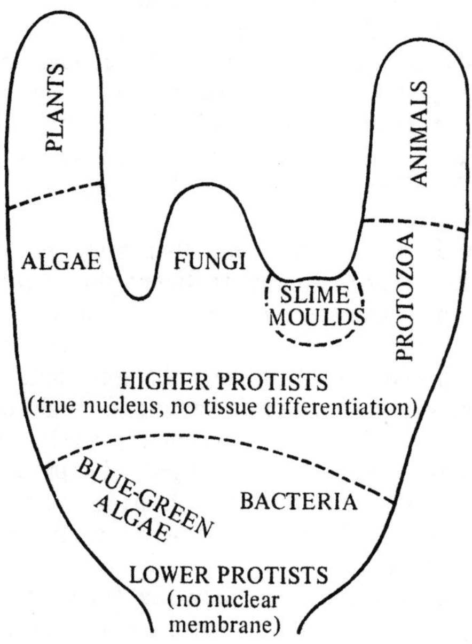

Fig. 1.1 *Diagram showing the interrelationships between plants, animals and protists*

avoids the term "microbe" or its equivalents altogether, shows the relationships of protists with other forms of life.

Cells

Basic Structure

We have already seen that cells may themselves constitute organisms, or they may be units from which organisms are built up. We have also seen that they may contain a clearly defined nucleus or nuclei, or merely nuclear material with no apparent organization. In addition, the cells of some types of organisms are surrounded by a cell wall of cellulose or some other material, which is more or less rigid and in some cases is thick, accounting for a considerable proportion (even up to 50%) of the dry weight of the cell; while other cells (for example the protozoan *Amoeba* and those curious anomalies the slime moulds) have no cell wall. In these latter cases, the protoplasm or cytoplasm of the cell is held together merely by a flexible membrane, the plasma membrane or cytoplasmic membrane – a structure which in fact also bounds the cytoplasm of walled cells.

The cytoplasmic membrane is a vital barrier to the cell, for it selectively absorbs certain substances from the environment – *i.e.*, it is semipermeable – while retaining the soluble components of the cell. Certain solutes can diffuse through the cell membrane against a pronounced concentration gradient, showing that absorption can be an active process dependent on the metabolism of the cell. Though in general small molecules pass through cell membranes while large ones (polysaccharides, proteins etc.) do not, there are obvious exceptions: extracellular enzymes (see below) pass out through the membrane, while the sulphur-oxidizing bacteria which survive highly acidic conditions must possess a mechanism for keeping out hydrogen ions.

The cell wall, when present, appears in most cases to be freely permeable to all solutes.

Cells occur in a great variety of shapes and sizes, as will be seen later.

Inside the cell various structures or organelles may be detectable, such as mitochondria (sites of respiratory processes), ribosomes (sites of protein synthesis), chlorophyll-containing chloroplasts (sites of photosynthesis, and, like nuclei, absent from blue-green algae in which the chlorophyll appears to be distributed uniformly), and – again absent from the lower protists – vacuoles containing cell sap. Because of this diversity in cell structure, it is unfruitful to attempt to describe or draw a "typical" cell.

When a cell divides into two, the daughter cells may separate, giving rise to two and eventually a large population of unicellular organisms, or they may remain attached to each other. If cell division occurs only in one plane, a chain, or hypha, or filament of cells will arise, and this is the simplest form of multicellular organism; but cell division may occur in two or three planes at right angles, resulting in sheets or solid masses of cells.

How cells obtain nutriment in order to grow, and exactly what we mean by growth, we shall see presently; but first we will take a brief look at the types of chemical substance present in the living cell.

Composition of Cells: Carbohydrates and Lipids

The most important constituents of microbes, quantitatively speaking, are carbohydrates, lipids, proteins and the non-proteinaceous nitrogenous substances found for example in nuclei.

Of the carbohydrates, polysaccharides are probably universal constituents of cell walls: for instance, cellulose occurs in the green algae, chitin in the filamentous fungi. However, lipid and protein are also found in the cell walls of some microbes. Other polysaccharides such as starch and glycogen are commonly found as storage material in microbes. Lower molecular weight carbohydrates (sugars) also occur, though not usually in significant amounts: they are important *metabolites* which, having been absorbed into the cell or arisen by hydrolysis of insoluble storage carbohydrate, are rapidly metabolized to yield energy or combined with other substances to form, for example, nucleotides.

Lipids are also frequently found as storage material, for which their high degree of reduction makes them particularly suitable. Droplets of fat or oil may often be demonstrated in microbial cells by suitable staining techniques. Another role of lipids and phospholipids, along with proteins, is in the construction of the cytoplasmic membrane.

Proteins; Enzymes

Proteins, of which the cell's cytoplasm is largely composed, are of particular interest to us. They are formed by the condensation of amino acids (general formula $R.CHNH_2.COOH$) into chains, and usually have a molecular weight in the region 20 000–100 000. Many of them, either as pure proteins or joined to a non-proteinaceous "prosthetic group" and then known as conjugated proteins, are the biological catalysts, *enzymes*, without which few biochemical transformations would be possible.

The biochemical processes of living things are collectively known as *metabolism*, and the substances on which enzymes act are called their *substrates*. As with ordinary catalysts, enzymes are not used up in the reaction they catalyse; nor, since they contribute no energy to the system, do they affect the position of the equilibrium finally reached. Their effect is to alter the reaction rate, sometimes speeding it up from virtually an infinitely slow rate; and they can catalyse a reversible reaction in either direction. The direction in which a given reaction actually proceeds in the cell will depend on such things as concentration of substrate and product, or perhaps redox potential, and hence on the activities of other enzymes – an example of the co-ordination and control which exists inside the living cell.

Two examples will serve to illustrate the efficiency of enzymes as catalysts. Cane sugar is a common energy source for microbes, and to hydrolyse it to glucose and fructose by chemical means would entail treating it with fairly strong acid at about boiling point. The enzyme invertase will however hydrolyse cane sugar very rapidly under neutral conditions at room temperature. The second example is the reaction of one of the most active enzymes known, catalase, one molecule of which is said to be capable of breaking down forty thousand molecules of hydrogen peroxide per second. This example also serves to draw our attention to the fact that the substrates of enzymes are not always organic.

Some enzymes require the presence of activators (*e.g.*, certain anions or cations) to enable them to function; some others require coenzymes (of which more below) to donate or receive certain atoms or groups. Coenzymes are different from prosthetic groups in that they can be handed around from one enzyme to another, donating or accepting their appropriate group in different reactions. Most enzymes are probably water-soluble, though some are associated with particulate matter in the cell. Others appear to be constituents of cytoplasmic membranes, while a few (the extracellular enzymes) pass outside the cell into the environment and hydrolyse insoluble material giving smaller soluble molecules that can diffuse into the cell. The mechanism by which the large extracellular enzyme molecules pass out of the cell is still something of a mystery.

A given enzyme can catalyse only one type of chemical reaction, and in fact many enzymes have such high specificity that they will catalyse one reaction only. The reason appears to be that there is an active group, or zone, on an enzyme molecule that actually "fits" its substrate, and enables the enzyme to form a complex with the substrate before acting on it.

Nucleotides

Certain heterocyclic nitrogenous bases (principally purines and pyrimidines) when combined with simple sugars and one or more phosphate groups constitute *nucleotides*. Some of these act as coenzymes, or carriers of reducing power (electrons) and high-energy phosphate groups between different enzymic reactions. Polymerized nucleotides constitute the well-known *nucleic acids* DNA and RNA. DNA is the material in the nucleus (or its equivalent in the lower protists) which carries the instructions for the organism's activities throughout its life cycle, while RNA is concerned with conveying this information from DNA to the sites of protein synthesis.

Other Soluble Constituents

The soluble fraction of cells contains besides various proteins, nucleotides etc.

a pool of metabolites of low molecular weight such as carboxylic acids, amino acids and sugars which though transient are sometimes present in detectable amounts in an extract of disrupted cells. In addition, certain species contain special substances such as pigments, which may have no obvious function. In some cases soluble intermediate metabolites are produced in abnormally large quantities and leak out of the cells: for example, species of the fungus *Aspergillus* produce citric acid in gross excess of their requirements. In other cases, substances such as organic acids arise as an end-product of a special growth reaction and are eliminated from the cell. Such excreted acids have an important role in corrosion.

Inorganic ions are also present in all cells. Some of these are required for enzymic reactions, as mentioned above, while others appear to be present in order to preserve ionic equilibrium between the interior of the cell and the environment.

* * *

Most or all of the constituents of living cells may be extracted and obtained in a pure or relatively pure condition; many can be crystallized. A mere examination of this collection of chemical substances gives us little real insight into the cell and its workings; and an analogy has been drawn with a visitor from another planet who, shown the pieces of a dismantled clock, would find it difficult to guess how they fitted together and impossible to guess the function of the device so formed.

The Elements Essential to Life

From the foregoing rather superficial survey of the constituents of the cells of microbes (and other living things), a list of the elements essential to life may be compiled. This list, below, will be seen to contain in parentheses a number of elements which are required only in relatively small, or even minute, amounts, but which are nevertheless vital for such purposes as the synthesis or proper functioning of certain enzymes, coenzymes etc.

C H O N S P (K Mg Ca Fe Cu Co Zn Mo Mn) .

This list is not exhaustive: only a very minute proportion of the known microbes have had their requirements studied. Some blue-green algae require sodium; diatoms (another type of algae) have siliceous cell walls; the presence of the chloride ion is probably essential for many marine micro-organisms.

We will now deal in turn with the classes of micro-organisms of which we are going to meet representatives in the chapters that follow.

Bacteria

As with many of the groups of microbes, the classification of bacteria is not only extremely complex but the subject of some controversy. The bacteria that interest us, though, are perhaps best regarded as all belonging to the *true bacteria* (eubacteria), a group characterized by possessing thick, rigid cell walls, dividing by binary fission (one cell dividing into two equal daughter cells), and, if motile, being so by means of one or more *flagella*.

The Unicellular Eubacteria

Most of the well-known bacteria are strictly speaking unicellular, though in many cases they occur in pairs, clusters or chains owing to incomplete separation after cell division. These aggregates are by no means "essential", however: each cell is autonomous and exists completely independently of its neighbours. This can be shown by mechanically breaking up the chains or clusters: the separated cells remain viable and normal.

Unicellular bacteria are of three basic shapes: spherical (called cocci), rod-like (bacilli) and curved rods (vibrios; or, if longer, with one or more complete turns of a helix, spirilla). There is a considerable variation in size, both between different species and in the same species under different conditions; and some bacilli are customarily so short as to be almost indistinguishable from cocci. A particularly large rod form, *Bacillus megaterium*, has dimensions 9–10 μ (microns) $\times$ 1·5 μ; *Haemophilus influenzae* is about 1 μ $\times$ 0·3 μ. Fig. 1.2 shows various types of unicellular bacteria; others are illustrated in Fig. 3.1.

Electron microscopy shows that eubacteria have little definite internal structure compared with the higher protists; but many of them (though never cocci) have whip-like outgrowths, the *flagella*, originating beneath the cell wall, and which can be clearly seen under the electron microscope and even under the light microscope with special staining techniques. The number and position of these flagella on a bacterial cell is often a help in identifying the organism: there may be one, or a tuft, or a large number distributed uniformly over the surface of the cell, or other arrangements.

The cells of all types of unicellular bacteria may be surrounded by gelatinous polysaccharide or polypeptide material constituting a *capsule*. Sometimes a more diffuse and bulky slime is produced, which can contribute for example to the blocking of drains. Such a slime producer is *Pseudomonas aeruginosa*.

A remarkable structure peculiar to the unicellular eubacteria is the *endospore*. This is a roughly spherical, dense, highly refractile body with low water content that is formed within the bacterial cell under certain conditions. Only one is formed per cell, and at a position characteristic of the particular species. It may have a greater diameter than the width of the vegetative cell, causing the organism to bulge. The only bacteria in which endospores occur are species of the rod-like genera *Bacillus* and *Clostridium*, rods and vibrios belonging to the genus *Desulfotomaculum*, and one species of coccus.

The remarkable feature of a bacterial endospore is its ability to resist

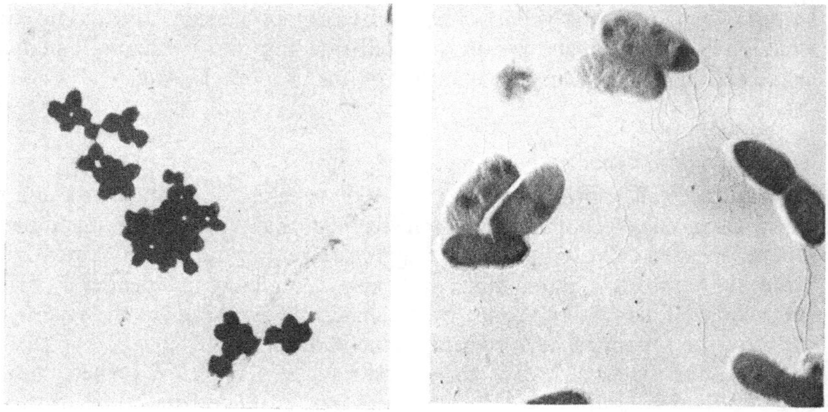

Fig. 1.2 *Electron micrographs of (left)* Micrococcus *sp.* (× 1650) *and (right)* Pseudomonas *sp.* (× 6250), *two types of bacteria encountered in infected emulsions*

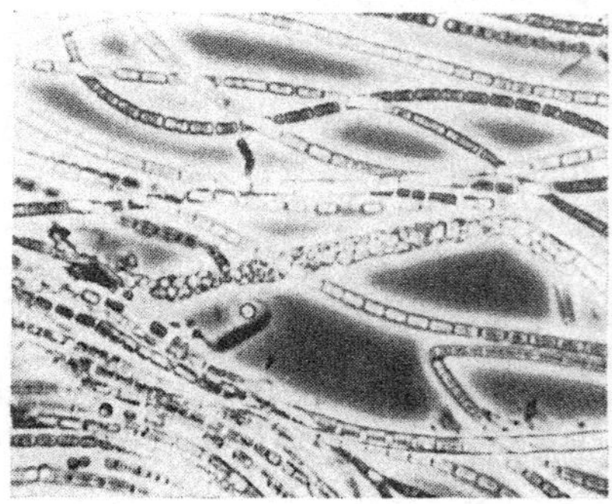

Fig. 1.3 *Electron micrograph of* Gallionella, *a stalked iron bacterium* (× 32 000)
Fig. 1.4 *Photomicrograph of* Sphaerotilus, *a filamentous iron bacterium* (× 700)

physical and chemical conditions greatly in excess of what would be lethal to the vegetative form of the same species, and to germinate to form a vegetative cell when conditions have returned to normal.

Certain unicellular eubacteria are stalked. One of the iron bacteria, *Gallionella*, consists of twisted, ribbon-like stalks each surmounted by a single curved cell. The twisted stalks, in reality merely a non-living excretion from the bacterium, were themselves formerly believed to be chains of microbes owing to their beaded appearance. Fig. 1.3 shows a *Gallionella* cell.

Filamentous Eubacteria

Other iron bacteria, though usually regarded as true bacteria, are filamentous in the sense that their cells occur in chains permanently embedded in an initially delicate organic sheath. An example is *Sphaerotilus* (see Fig. 1.4). Because of its metabolic activity, dealt with in more detail under *Nutrition* below, its sheath eventually becomes heavily impregnated with yellow-brown ferric hydroxide and thickens considerably so that the organism forms sizeable tufts (the so-called "sewage fungus" of effluent channels). The ferric hydroxide-impregnated form of *Sphaerotilus* was originally regarded as a separate organism, and variously called *Leptothrix* or *Crenothrix*.

Reproduction of these filamentous forms is by the production of motile rod-like "swarmers" (the filamentous bacteria are otherwise non-motile) which escape from the open ends of the sheaths.

Mycelial Eubacteria (Actinomycetes)

These organisms have a mycelial structure without cross-walls, and so superficially resemble fungi (see below), but the hyphae are very narrow – the same order of thickness as the unicellular bacteria – and an actinomycete is best regarded as a vastly elongated, branched bacillus. They are common inhabitants of soil, and have been said to give damp soil its characteristic smell. Old mycelium frequently fragments into small coccoid or rod-like bodies indistinguishable from unicellular bacteria. In some species, reproduction is by formation and subsequent scattering of spherical cells, *conidia*, at the tips of hyphae.

* * *

A few bacteria are photosynthetic, though their photosynthetic mechanism differs somewhat from that in the algae and higher plants.

Algae

The algae are chlorophyll-containing organisms (plus a few colourless forms which are obviously derived from pigmented ones) possessing the same photosynthetic mechanism as higher plants which thus distinguishes them from the photosynthetic bacteria. They are thalloid; that is, they have no true tissues despite the enormous size of some of the marine algae. They are divided into a number of groups on the basis of the pigments they contain (chlorophylls and accessory pigments), type of flagella when present, life cycle and type of reproduction, etc.

Blue-green Algae

This is the only group of algae that possesses no nuclei or chloroplasts, and hence belongs to the lower protists. The bluish colour is due to the presence of coloured conjugated proteins called phycobilins. They are all small algae, coccoid or filamentous (usually unbranched); some are of a size comparable with largish bacteria. Growth rates tend to be higher than those of other algae.

The cell wall consists of an inner cellulose layer, then a pectic layer, then outside this again a layer of mucilage. This outermost layer may be very thick, embedding many organisms in what appears to be a common gelatinous sheath. Heavy slime producers are frequently responsible for the blockage of industrial water systems (see Chapter 4), and for the development of "blooms" covering lakes and ponds which set up a chain of events resulting in the death by suffocation of fish.

Reproduction of the blue-green algae is usually by binary fission, in both unicellular and filamentous forms, though some types produce spores. Sexual reproduction does not appear to occur, nor are there any motile stages. Filamentous blue-green algae do nevertheless show slow movements, attributed to the secretion of mucilage.

Several species fix atmospheric nitrogen – a rare property amongst living things, bestowing an obvious advantage in enabling infertile situations to be colonized. Another highly unusual feature is the ability of several types to thrive at what we would regard as "lethal" temperatures, *i.e.*, they are thermophilic. This is dealt with more fully under *Growth*, below.

Green Algae

The green algae are probably the most recent to have evolved, and many people regard them as the closest relatives of the higher plants. They contain the same chlorophylls; and their cell walls, which consist of cellulose and pectin, are basically similar to those of higher plants. A nucleus and other organelles typical of the higher protists occur, and older cells contain a vacuole filled with cell sap. The photosynthetic pigments are contained in chloroplasts; but unlike those of higher plants, which are simple lenticular bodies, those of the green algae are sometimes very elaborate and beautiful. A famous example is the common filamentous pond alga *Spirogyra*, with its helical chloroplasts arranged round the periphery of the cell.

Green algae show a wide range of structure from unicells (sometimes motile), through colonial forms, filamentous and sheet-like forms, to the stoneworts (*Chara*), reaching 3–4 ft in length and looking superficially like higher plants. Some authorities in fact place the stoneworts in a separate group, though they have clear affinities with the green algae.

In as diverse a group as this, it is not surprising to find a wide range of methods of reproduction. These include simple cell division, formation of spores (motile or non-motile), and various forms of sexual reproduction.

Diatoms

This is a group of algae that are very abundant in fresh and salt water (both as planktonic and bottom-dwelling forms) and in the soil, and includes some of the most beautiful of microscopic organisms. Most are unicellular; many have strange and elaborate geometrical forms with intricate sculpturing on the cell wall which is heavily impregnated with silicon. The cell wall is itself an unusual and interesting structure: it consists of two halves fitting one inside the other like a box and its lid; and, to press the analogy further, a diatom cell, like a pill box or date box, has quite a different appearance according to the direction from which it is viewed. Often there is a furrow in the wall through which mucilage is secreted. This extrusion of mucilage propels the diatom slowly along a solid surface.

Diatom cells possess a nucleus, and one or more chloroplasts. Accessory pigments often give them a brownish colour. Vegetative reproduction occurs by nuclear division followed by separation of the two halves of the cell wall so that each contains cytoplasm and a nucleus. A new half-wall then forms on each daughter cell. Sexual reproduction also occurs.

Common diatoms are illustrated in Fig. 1.5; green and blue-green algae are also shown in this figure and in Fig. 1.6.

Fungi

Moulds

The fungi are a group of mainly filamentous organisms, always containing nuclei. The filaments or *hyphae* are branched, and a tangled mass of hyphae is referred to collectively as *mycelium*. The hyphae themselves either have no cross-walls, being merely tubes containing cytoplasm with many nuclei, or they have cross-walls (septa) at intervals, each with a pore in the centre so that there is still cytoplasmic continuity along the hypha. The compartments in a septate hypha are thus not cells: nuclei can squeeze through the pores in the septa so that the individual compartments do not necessarily contain one or even a constant number of nuclei. Therefore there are grounds for regarding

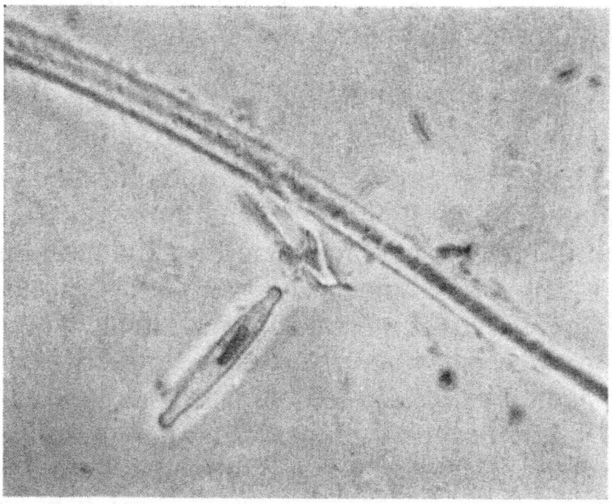

Fig. 1.5 *Diatoms and a filamentous blue-green alga from a mixed growth of algae on cooling tower timbers* (× 500)

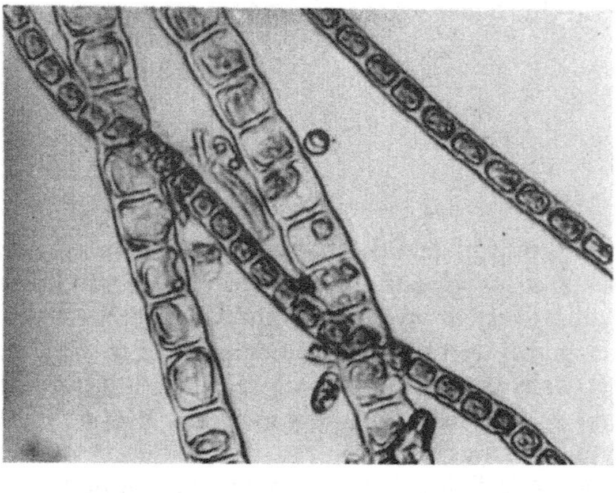

Fig. 1.6 *Filamentous green algae from the same source as in Fig.* 1.5 (× 460)

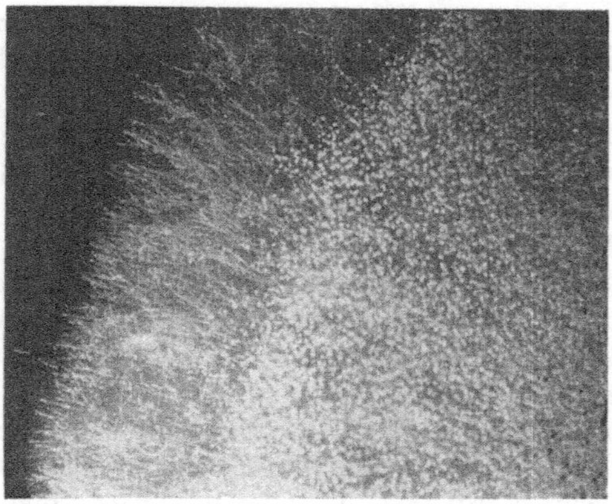

Fig. 1.7 *The margin of a colony of a mould,* Aspergillus *sp., growing on the surface of solid medium.* Black (ripened) conidial heads can be seen at the bottom of the picture

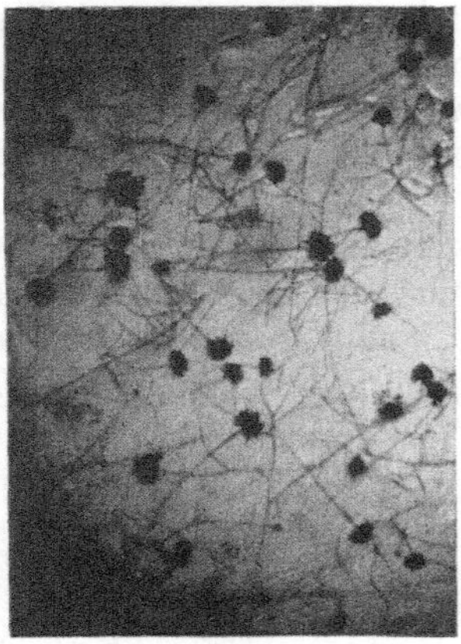

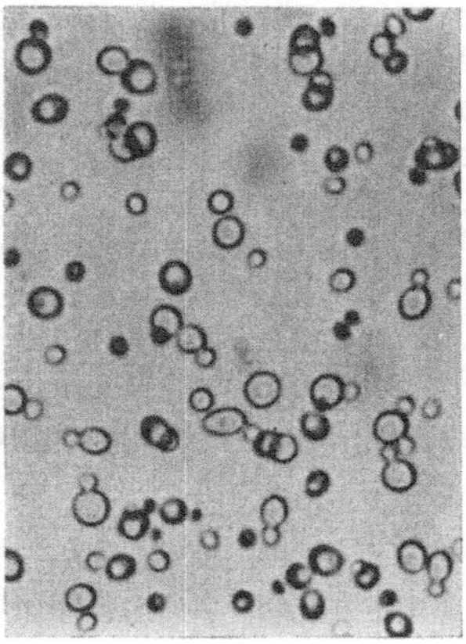

Fig. 1.8 *Hyphae and conidial heads of* Aspergillus *sp.*

Fig. 1.9 *Yeast cells* (Candida *sp.*), *showing reproduction by budding* (× 350)

such a mycelium, originating from a single spore, as a single-celled multinucleate organism or "coenocyte", whether or not it is septate.

The diameter of fungal hyphae varies considerably, even in a single tuft of mycelium, but they are always much thicker than the hyphae of actinomycetes: the narrowest have a width of about 3–4 μ, while 20–25 μ is not uncommon. Fungi are never photosynthetic.

There are thousands of species of moulds (about 15 000 of Ascomycetes alone); and, as with the bacteria, classification is something of a problem. Sexual reproduction (involving fusion of nuclei) is common, and in addition all fungi have some method of asexual reproduction. Figs. 1.7 and 1.8 show asexual reproductive bodies in an Ascomycete.

In the largest class of moulds, the Ascomycetes, sexual reproduction results in the formation of a characteristic sac-like body containing four, eight or occasionally sixteen ascospores. This class has septate mycelium, as has the class Basidiomycetes which includes the fungi that produce mushrooms and toadstools. In a large number of septate fungi, though, a sexual process either does not exist or has not yet been observed; they are therefore provisionally classed as Fungi Imperfecti though undoubtedly many of them have affinities with the Ascomycetes. The moulds mentioned in this book belong either to the Ascomycetes or the Fungi Imperfecti.

Yeasts

Yeasts normally appear under the microscope as separate cells, which may be spherical, ovoid, elongated or other shapes. During rapid growth they may form chains of elongated cells. Small yeast cells may be 3 μ in diameter; large elongated cells of some species reach 20 μ or more in length. Growth of yeasts on solid substrata forms typical moist, shiny colonies. Some species form true mycelium under certain conditions; this, together with the occasional formation of an ascus-like body, reveals the affinity between the yeasts and the Ascomycetes.

The yeasts have a unique mode of vegetative reproduction known as *budding*. A cell can form one or more projections or bulges, which eventually grow to about the same size as the parent cell when constrictions form and the daughter cell or cells break off. Budding cells can be seen in Fig. 1.9. In addition, certain yeasts reproduce by binary fission.

The Success and Ubiquity of Microbes

We have seen that microbes are all relatively simple in structure and organization except at a cellular level. It is quite another matter, though, to dismiss them as "primitive": this might suggest that they are ancient, experimental prototypes of living things, lacking refinements and with built-in defects.

Ancient they may be, compared with the higher animals and man; yet this has given them time to adapt, and adapt supremely well, to almost every conceivable habitat on land and in the water. The main advances we higher animals have made include carrying our own stable environment around inside us and being able to move over large distances in search of better conditions for survival. Microbes, because of their small size, have a large surface/volume ratio and hence are easily and rapidly affected by an unfavourable change in their environment; thus they cannot in general maintain such stable internal conditions. On the other hand their simplicity of structure and the rapidity with which they can absorb nutrients on a body-weight basis enables them to multiply exceedingly fast compared with higher animals. This in turn means that genetic mutations can arise with much greater frequency so that there is a good chance of new and resistant forms arising before environmental changes have wiped out a microbial population. Furthermore, the slowness of microbial motility, when present, is compensated by their often highly efficient methods of spore dispersal. The advantages of the resistant bacterial endospore are obvious; but the spores of fungi, etc., though without enhanced resistance, can be carried great distances by water or air currents and some are certain to find suitable conditions for germination.

Summing up, many microbes have over the geological time-scale become adapted to highly specialized and bizarre environments, where they face no competition from other organisms. Others are as well adapted to their more commonplace environments, and thus all can be regarded as simple but highly successful forms of life. We higher animals have achieved success in a different way, in that there is less sacrifice of *individuals* in the process of adaptation – instead, each individual is highly adaptable.

The result is that the earth's crust is teeming with microbes of every kind, either as actively metabolizing, growing cells (on the earth's surface, at considerable distances below, and right down to the bottom of the ocean), or as dormant forms in very dry or cold conditions, floating in the air etc. As will be seen, microbe-free conditions do not just arise: they have to be striven hard for.

Growth

We have already noted that microbes, because of their great surface/volume ratio, can absorb dissolved material very rapidly. Under favourable conditions – a plentiful supply of all its essential nutrients, though not in a high enough concentration to be toxic – a bacterium for example can add sufficiently to the materials that form its cell to be able to divide into two organisms by binary fission in as short a period as twenty minutes after the previous division or even less. In other words, chemical synthesis in a unicellular organism under such optimum conditions results in a balanced increase of its cell components directed towards the production of another cell. We may make this the basis for our definition of growth. The outcome of this process is that a unicellular organism will produce an exponentially increasing number of unicells, a multicellular organism will acquire more cells and hence increase in size, while a fungal mycelium or algal filament will grow in length and soon become a tangled mass. Certain forms of increase in cell size not accompanied by cell division – for example, swelling of a cell when immersed in pure water, or excessive accumulation of food-reserve material in response to deteriorating external conditions – are not regarded as true growth by biologists.

Exponential growth of a unicellular organism – binary fission of all cells at a constant rate – is a formidable thing. A single bacillus of quite normal dimensions, dividing this way every half hour for 36 hours, would cover the whole surface of the earth with bacilli to a depth of about a metre! This could never happen, of course, because firstly the earth (assuming it were flat) would need to be covered with a suitable nutrient medium to a depth of anything up to 100 metres, and secondly, for this rate of growth to be maintained at high microbial population density in such a deep medium we would have to devise a way of efficiently mixing this growing culture to keep all the organisms supplied with nutrients, and of getting enormous quantities of dissolved oxygen into the culture medium. Fermentation technologists, in fact, are much preoccupied with these very problems when growing their modest-sized fungal or bacterial cultures for producing citric acid, antibiotics etc. in fermenters of capacity a few thousand litres.

In the type of microbial growth (called *batch culture*) that I have envisaged – either that covering the earth, or in a fermenter or for that matter in a small glass vessel on the laboratory bench – there is obviously a steady drop in nutrient concentrations in the medium as the microbial population increases. Eventually the growth rate will become limited by a shortage of one of the nutrients, and soon afterwards growth will stop altogether.

Another method of growing microbes is called *continuous culture*, and consists basically of running fresh medium in at a steady rate and allowing excess microbial culture to run off. In natural conditions, with which we are going to be largely concerned, the growth of microbes frequently resembles that in a rather severely nutrient-limited continuous culture more than that in a batch culture. The reason is that in natural situations there is often a throughput of water (*i.e.*, it is a river, or soil through which there is water movement – and all natural water contains some nutrients), or alternatively there is a slow release of soluble nutrients held as complexes by solid particles or produced by microbial hydrolysis of insoluble organic matter. As already hinted at, the rate of ingress of oxygen often severely limits growth rates. Thus the rapid exponential growth we are able to produce in cultures in the laboratory is undoubtedly the exception rather than the rule in nature.

All living things have an optimum temperature for growth. The more highly evolved vertebrates have a mechanism for regulating their internal temperature; other organisms are at approximately the same temperature as their surroundings. Probably most microbes have an optimum temperature – *i.e.*, one at which the rate of growth is most rapid – somewhere in the temperature range of their normal habitat. If this habitat has an air temperature typical of temperate or warmer climates, such organisms are called *mesophiles*. In arctic regions and in the deeper waters of the oceans, many *psychrophilic* microbes exist, with an optimum temperature in the region 4–10 °C. At the other end of the scale are *thermophiles* whose optimum lies at or above about 55 °C; and this group contains some remarkable microbes including a bacterium reported to grow at 92 °C and several blue-green algae which can withstand 75 °C. Thermophiles are, curiously, often found in situations of normal temperature where presumably they are either lying dormant or only growing very slowly.

Nutrition

We now know which elements are essential to life, and must therefore be available as solutes or gases to the microbes we are considering. The form in which these elements can be utilized, though, varies greatly from one type of micro-organism to another. It is useful to consider all living things as clearly bounded zones in which the state of reduction and complexity is markedly higher than that surrounding them: the organism thus has to build up its essential organic molecules from dissimilar material in the environment, and in order to do this it must be able to perform work on this environmental material. All living things require water, though some have highly remarkable powers of conserving it and can therefore live in extremely arid places. As to the other nutrients, we will deal now with some of the principal ways in which microbes obtain them.

Photosynthetic Organisms

Carbon, the backbone material of all organic compounds, is present (and is becoming ever more present) in our atmosphere as its fully oxidized form, CO_2. This cannot be incorporated (except to a very limited extent) by living things unless they have a means of tapping a source of energy with which to reduce it. The commonest source of energy for this purpose is visible light, which is used by the photosynthetic organisms already mentioned with the catalytic action of chlorophyll. The hydrogen for the reduction comes from water, and the primary products of the reaction are oxygen and a sugar phosphate. Such organisms are often called autotrophs, but are more accurately described as *photolithotrophs;* and they can, with very few exceptions that we know of, synthesize all their organic cell constituents using carbon from CO_2.

Photosynthetic organisms thus need only CO_2, water, sunlight or artificial light, and a few simple mineral salts to provide such ions as NH_4^+ or NO_3^-, K^+, Mg^{2+}, Ca^{2+}, SO_4^{2-}, PO_4^{3-} etc. All of these mineral ions occur, at least in traces and often abundantly, in natural waters and soils; but in any case, many blue-green algae, as we have already seen, do not need "fixed" nitrogen but can use atmospheric nitrogen.

Other Methods of Fixing Carbon Dioxide

A few bacteria have acquired the ability to perform certain inorganic oxidations and to couple these to the reduction of CO_2 in an analogous way to photosynthesis. These organisms are called *chemolithotrophs*. One such group is the sulphur-oxidizing bacteria, which oxidize sulphur or its compounds such as sulphides to sulphates (this is described in more detail in Chapter 4).

Another group is the iron bacteria. Natural waters, as well as industrial waters in contact with iron or steel, frequently contain ferrous ions in solution, and the iron bacteria can obtain energy for CO_2 fixation by oxidation of these to ferr*ic* ions resulting in the troublesome formation of ferric hydroxide.

Thus organisms which obtain their energy for growth reactions chemolithotropically or phototropically utilize CO_2 as their carbon source: they are all *autotrophs*.

Yet another oxidizable inorganic substance is hydrogen. Elemental hydrogen is not particularly abundant on the surface of the earth, though it does occur for example on the surfaces of metals under certain conditions; and Chapter 3 deals with bacteria which can oxidize it.

* * *

Photolithotrophs and chemolithotrophs, then, are the *primary producers* of the living world; all other microbes, and all animals, depend for their carbon and energy ultimately on the cell material of these primary producers. Creatures that directly devour these CO_2 fixers are outside the scope of this book, but all the other microbes we have considered depend on organic substances, either present already in a soluble form in their environment or solubilized by the microbe's own extracellular enzymes, and which could not have arisen but for the primary producers. It follows that the energy fixed in the organic material in the biosphere, except for the small contribution of the chemolithotrophs, all derives ultimately from the sun.

The first colonizers of "clean" water – water containing only traces of mineral salts – are therefore the algae. When these die by falling into unilluminated places or for any other reason, and decompose, organic material is then available for the vast number of chemo-organotrophic microbes (see below) to feed on. Standing bodies of water seldom stay clean for long – organic debris soon accumulates in them – and so the initial algal stage is often unnecessary for the development of a huge microbial population.

Oxidative Assimilation of Organic Substances
Most non-photosynthetic microbes obtain both their cell carbon, and their energy for building it into complex molecules, from soluble, low molecular weight organic substances by a process called *oxidative assimilation*. Fig. 1.10 is a grossly oversimplified way of illustrating this. It will be seen that gaseous

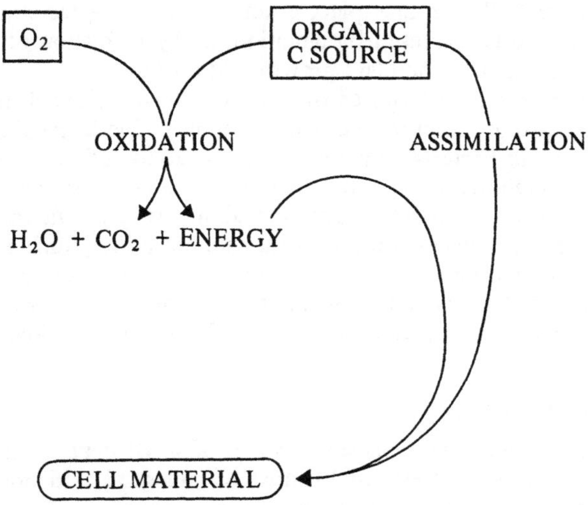

Fig. 1.10 *Simplified scheme to illustrate oxidative assimilation*

oxygen is required; we will look at this aspect in a little more detail soon.

The organic substrate for this type of growth is commonly a monosaccharide such as glucose, or other soluble carbohydrates, or lower carboxylic acids, alcohols etc. The utilization of organic substances as energy sources is called *chemo-organotrophy*, and their utilization as carbon sources is called *heterotrophy*. Almost all heterotrophs are chemo-organotrophic, since they use processes like the oxidative assimilation pathway outlined above. Many heterotrophs can obtain all the other elements they require from the types of inorganic ions listed under *Photosynthetic Organisms*. Other heterotrophic microbes are much more exacting: they may require in their growth medium one or more specific organic substances that they are unable, by an evolutionary accident, to synthesize for themselves.

Certain microbes require, or at least they can manage very well with, rather bizarre carbon-and-energy sources. Some organisms that we will meet in the following chapters are of interest to us for this very reason. As we have seen, insoluble substances – structural materials – are not immune to attack. Cellulose (a constituent of fabrics used for protective wraps around pipes) and lignin (a constituent of wood) are hydrolysable by certain bacteria and fungi. Hydrocarbons – gaseous, liquid and solid – can serve as substrates for growth. The utilization of liquid hydrocarbons is perhaps especially surprising; however, the lower ones have a slight miscibility with water (which is essential to life), and oxygen is slightly soluble in them. Water, in fact, seems to be no great problem to the hydrocarbon utilizers: it readily appears to become entrained in fuel oils and settles out on tank bottoms and other low spots. Growth will then take place at an oil/water interface; and if the area of this interface is increased by agitation, or by formation of an emulsion, much more rapid and abundant growth will result.

Other microbes have acquired the ability to metabolize substances that we normally regard as lethal to microbes. Phenol is "biodegradable" by certain yeasts, fungi and bacteria – though we can make use of this in the treatment of phenol-containing industrial effluents. It should also be mentioned in passing that certain microbes have the ability to grow in the presence of "toxic" concentrations of substances they do not utilize, such as heavy metal ions, sulphide, sulphuric acid etc., or to adapt rather rapidly to toxic substances employed as disinfectants against them (see below) owing to the high rate of production of mutations in a heavy microbial population.

Cellular Oxidations and Reductions

Biochemical reactions take place in a series of small steps, each catalysed by a different enzyme. The essential feature of biological oxidation is not combination of the substrate with oxygen but the stepwise removal of electrons from it and their acceptance by various carriers which constitute the cell's electron transport system or reservoir of reducing power. There must, of course, be a *terminal electron acceptor* for this energy-yielding process, and this overall process is what we know as respiration.

In the vast majority of living things, including most of the chemo-organotrophic microbes and the photosynthesizers when they are in the dark, the substrate (electron donor) for respiration is an organic substance of some kind and the terminal electron acceptor is atmospheric oxygen which is reduced to water. This constitutes *aerobic respiration.*

As we might expect, though, microbes have as it were explored the alternative possibilities, and two alternatives to aerobic respiration have evolved. One is *fermentation*, in the strict, biochemist's sense (to the biochemical engineer, a "fermentation" is any type of microbial growth resulting in an organic end-product apart from the cells themselves). A true fermentation is an energy-yielding process in which the electron *acceptor* as well as the electron donor is an organic substance, and in which therefore atmospheric oxygen does not participate. The alcohol that brewer's yeast produces by reduction of pyruvic acid, one of its own intermediate products of sugar breakdown, has functioned as terminal electron acceptor; well-aerated yeast cultures do not produce alcohol but utilize oxygen.

Another variant of respiration is *anaerobic respiration* or the use of inorganic substances as electron acceptors. Some bacteria are able to reduce nitrate to nitrite or nitrogen, while the notorious sulphate-reducing bacteria (see Chapter 3) reduce sulphate to sulphide. There is, incidentally, an apparent paradox here. We have seen that certain bacteria oxidize sulphur and some of its compounds to sulphate while certain others perform the reverse reaction (the consequences of both types working together are described in later chapters); yet these, like all living things, are striving for the same goal: the production of more individuals in their own likeness. The sulphur oxidizers couple their inorganic oxidation with CO_2 reduction; the sulphate reducers couple their reduction with oxidation of an organic substance. Clearly the overall reaction in both cases must be exergonic.

Many microbes possessing these alternative mechanisms to aerobic respiration are totally unable to grow in the presence of oxygen. This may be because they lack the necessary enzymes to use molecular oxygen, such as catalase, already mentioned, that breaks down toxic hydrogen peroxide that tends to arise during oxygen reduction. In the case of the sulphate-reducing bacteria, they will not grow if traces of oxygen are present since there cannot then exist a redox potential sufficiently negative for the reduction of sulphate to sulphide.

A few microbes grow best in the presence of low concentrations of oxygen; these are called *microaerophiles*.

Biochemical Oxygen Demand (B.O.D.); Microhabitats

It will be clear from this account of nutrition that water polluted by organic matter will support a heavy growth of chemo-organotrophic microbes which, unless the water is turbulent, will cause the dissolved oxygen to fall to a very low level or even to zero. The total oxygen uptake by a sample of polluted water can be determined by various methods in which an excess of oxygen is made available: its uptake is then a measure of the amount of oxidizable organic matter initially present, that is, the degree of pollution, and is called the B.O.D.

When waters become depleted of oxygen in this way, by the activities of so-called "scavenging aerobes", conditions in the deeper layers and in the bottom mud become favourable for the growth of fermentative and putrefactive organisms (which produce a variety of organic end-products including acids, and often unpleasant odours) and of course the sulphate reducers.

In an analogous way, conditions can become favourable for the growth of anaerobes in special situations (microhabitats) which may be only a few millimetres from a plentiful supply of oxygen. A tuft of filamentous aerobic microbes growing on, say, a metallic surface can easily have such a high oxygen demand that none can reach the middle of the tuft, and here any embedded sulphate reducers will be able to grow.

Cultivation of Microbes

Suitable media for the laboratory culture of microbes are given in several books (see *Suggestions for Further Reading* at the end of this chapter). Photosynthetic organisms are easy to grow in the light in test tubes or flasks containing dilute solutions of mineral salts or soil extract, if a crude culture is all that is desired. For non-photosynthetic forms, unless we happen to know their nutritional requirements, we must err on the generous side and give them a rich medium containing substances such as meat or yeast extract, protein hydrolysate etc., rather than a glucose + mineral salts medium.

Rich media in particular have the disadvantage that most of the microbes lurking on the glassware, in the laboratory air etc. will be able to grow in it too. Hence the paramount importance in the microbiological laboratory of aseptic techniques. The rationale of sterilization is dealt with in the next section of this chapter; for now, suffice it to say that all growth vessels, pipettes, nutrient media etc. must be sterilized, preferably in an autoclave, and enclosed in some way so that they remain sterile. Non-absorbent cotton wool is a boon to the microbiologist: a vessel of medium plugged fairly tightly with it will remain sterile once autoclaved, and when the medium is deliberately inoculated with some micro-organisms enough oxygen will diffuse through the plug to support quite heavy growth.

Sometimes tubes or dishes of solid medium (ordinary nutrient medium of some kind, made into a gel by adding agar) are used. These have the advantage that the microbes, if spread thinly on to the surface, develop into separate colonies often with a characteristic appearance that helps in identifying them and recognizing the presence of contaminant organisms.

Certain specialized techniques for cultivating microbes of economic importance are described in the following chapters.

Sterilization and Allied Processes

Of as much importance as how to grow microbes is how to prevent them from growing, or in more general terms, how not to have microbes. This is a meaningless concept except in relation to a definite object, or volume of space, delimited by a microbe-proof boundary. Given such a closed system, we can take steps to kill the inevitable microbes already present, and to kill or mechanically remove the microbes in any stream of liquid or gas entering the system. (The cotton wool plugged small-scale culture vessel we have already mentioned is microbiologically such a closed system: it has been sterilized, probably by autoclaving, and although air can enter it, all microbes are filtered out of this air by the compacted cotton wool).

First, then, it might be appropriate to look at the main types of process that can be used to achieve "sterilization"; and then, to examine the term more closely.

Methods of sterilization can be conveniently divided into mechanical, physical and chemical.

Mechanical Sterilization

We have already met the cotton wool plug. Many fibrous materials, packed so as to avoid "channelling", can be used for filtering out microbes from a stream of air, provided the fibrous material in its container is initially sterilized, and it remains dry. (Films of moisture "conduct" microbes with great facility.)

Both liquids and gases can be freed from microbes by passage through a filter of pore size of the order of 1 μ: this is called an absolute filter and differs in principle from the packed-fibre filter in which the space between the fibres is very much greater than the size of microbes, and which collects the microbes electrostatically (another good reason for keeping it dry). Absolute filters easily become choked with particulate matter, vastly cutting down the flow rate: the liquid or gas is therefore passed through a coarse filter, or a series of filters, before reaching the microbial filter.

The two other types of sterilization are concerned not with preventing the entry of microbes, but with killing those present.

Physical Sterilization

The principal method of physical sterilization is by heat. Wet heat is much more effective than dry heat, evidently because proteins containing moisture easily coagulate (alter irreversibly) with heat, whereas dried-out proteins have a much greater heat stability and suffer mainly from slow direct oxidation. Hence bacterial spores, with their low water content, are more heat-resistant than vegetative forms.

It is difficult to prescribe a wet-heat treatment that will kill all forms of microbial life, one reason being that it is often doubtful whether every part of the system being heat-treated reaches the required temperature or remains at it for the whole of the desired holding time. Steam at 15 p.s.i.g. (corresponding to about 121 °C) for 20 minutes in an autoclave almost always achieves a complete kill; there has been an isolated report of certain bacterial spores surviving 120 °C for 35 minutes.

Intense dry heat such as a direct flame is naturally an efficient means of sterilization, though of limited usefulness. It is used in the microbiological laboratory for sterilizing the neck of a vessel after removing the cotton wool plug, sterilizing a wire loop before introducing it into a culture vessel to pick up some micro-organisms, etc.

Ultraviolet, gamma and electron beam irradiation can under certain conditions be successfully employed to kill microbes.

All the foregoing methods of sterilization have the advantage from the microbiologist's point of view – the disadvantage, from certain others – that conditions in the sterilized system are, or become immediately on cooling, suitable for the growth or at least survival of any microbes that may subsequently enter. This is broadly untrue of the next method we shall consider.

Chemical Sterilization and Inhibition

By now it has become imperative that we examine the meaning of some of the more useful terms used in the subject of microbial inhibition.

Sterilization is an absolute term, correctly applied only to a process which kills or removes all forms of life in a given system. *Disinfection* is a less specific term, implying a reduction in the number of microbes present, or the destruction of certain types that are producing an unwanted effect.

Physical or chemical sterilizing agents do not achieve their effects instantaneously; it is thus necessary to know the time that a given treatment (concentration, temperature, intensity of radiation) takes to effect a complete kill of a specified type of microbe. As we shall see, the efficacy of a chemical sterilizing agent may vary greatly according to the physical and chemical conditions obtaining in the site where it is employed.

Incidentally, since it is almost impossible actually to find unicellular microbes except in a given space of literally microscopic proportions, we cannot use any form of direct observation to ascertain whether they are present, and if so, whether they are alive or dead. We assume that a microbe is dead if it has lost the power to reproduce, as determined by putting it into, or letting it fall into, some liquid or solid nutrient medium and observing whether visible colonies or turbidity develop. Therefore we perform a growth test on what we hope is a representative sample of liquid or gas from our system, to show up the presence of viable organisms. A substance or treatment that achieves a complete kill by this criterion is said to be *bactericidal* (fungicidal; algaecidal). A complication arises, though: a substance correctly describable as a bactericide need not necessarily kill bacterial spores.

Most, if not all, substances that kill microbes at a certain concentration will at lower concentrations prevent them from growing or reproducing, while not actually killing them. If the substance is then removed or diluted, growth will start again. This is known as a *bacteriostatic* (fungistatic etc.) effect. For many purposes it is quite sufficient to achieve this state of bacteriostasis and save the trouble and expense of complete sterilization: most of the undesirable effects of microbes are due to their growth. For obvious reasons, this kind of treatment is only practicable in a "closed" system (most large-scale systems – for example recirculating cooling-water systems open to the atmosphere and subjected to make-up water and bleed-off – are only closed in a relative sense); and then only if the microbial population already present is not troublesome in a purely mechanical way, by coating heat-exchange surfaces, blocking up filters and nozzles etc. In fact it is essential initially to kill most of the microbes in a badly fouled system by a large "slug" of a disinfectant (this usually causes those adhering to surfaces to drop off or enables them to be washed away), and after removal of as much of the dead material as possible for the same reasons as above, to dose the system with a bacteriostatic concentration.

It may be necessary at some stage to change to another disinfectant: microbes, as has already been mentioned, can sometimes adapt to an inhibitory concentration of a given disinfectant and resume growth.

In the case of bacteriostatic agents (bacteriostats), then, it is only necessary to know the concentration required to inhibit the growth of a specified type of organism, with the same proviso as mentioned for sterilizing agents.

Nomenclature of Chemical Inhibitors: Recent Coinages

Recently the term *biocide* has been introduced into the literature, apparently to denote any chemical with which effective killing and subsequent control of undesirable microbes in an industrial situation has been accomplished. It would thus seem to mean the same as "disinfectant"; etymologically, it means killer of any or all forms of life. Clearly, then, it cannot be used in any exact sense – some cautious writers use the alternative form *biostat* – and though it is thus somewhat misleading, it perhaps has a useful function in denoting a substance that has been found, probably on an empirical basis, to be effective in controlling undesirable microbes on an industrial scale. It is certainly true that many of the substances listed as bactericides etc. as a result of laboratory studies could, for various reasons, never be employed successfully on an industrial scale.

Another new term which appears to be used in the same sense as biocide is *microbicide* (or microbiocide).

The properties sought for in a biocide are discussed in Chapter 4.

Factors Affecting Biocide Action

We have seen that organic matter inevitably accumulates in a system in which there is the possibility of algal growth, let alone that of the entry of organic debris such as dead leaves, and that this results in the development of a rich and varied microflora. The presence of organic matter also reduces the effectiveness of biocides, as was recognized long ago in the laboratory testing of disinfectants.

There are several reasons for this. Perhaps the most common is a direct chemical reaction between the biocide and the organic matter – either oxidation by the biocide, as happens for example with chlorine, or the formation of a complex which always proves to be less toxic than the original biocide, or formation of an insoluble compound. Some organic matter precipitated this way, and some types of dissolved organic matter, also appear to be able to form a protective layer on a microbial surface which enhances their resistance to biocides.

Phenol is reported to be more active in saline than in water, but on the other hand the presence of sodium chloride is known to reduce greatly the bacteriostatic power of quaternary ammonium compounds. There can be little doubt that other inorganic solutes affect biocide activity.

Oils, fats and greases depress the apparent activity of many biocides such as phenols, or indeed any substance with a high oil/water partition coefficient, simply because most of the biocide passes into the non-aqueous phase.

Biocides should not be mixed indiscriminately. Many surface-active agents, for example the quaternary ammonium compounds, are employed as biocides; these are cationic detergents and become inactive if mixed with anionic detergents. Yet low concentrations of surface-active agents may, by increasing the permeability of microbial cells, assist the action of certain other types of biocide.

As to physical conditions, the activity of most biocides is enhanced by rise in temperature: it commonly increases two- to fourfold for every 10 °C rise. But some substances such as reducing agents are not affected in activity by temperature, and a few (sodium hydroxide is one) decrease in activity with increasing temperature.

It is not always safe to assume that a concentration a little below the recommended dosage will suffice. The "dilution coefficient" has been determined for many of the standard disinfectants; the value for phenol is about 5, which means that halving its concentration multiplies the disinfecting time by 2^5, $\times$ 32. This effect will thus be especially important in a "slug" treatment designed to achieve a relatively complete kill during a short shut-down period.

A knowledge of these chemical and physical properties of a biocide can of course be of great help when trying to employ them to best advantage. Unfortunately such data are seldom available, and treatment then has to be based on often only one or two published accounts of successful procedures in other systems, in which the physical and chemical conditions and the type of infection were probably very different from those in the system needing treatment.

Biodeterioration

Much of this book is devoted to metallic corrosion by microbes; but this subject forms part of a larger topic, *biodeterioration*, the economic significance of which has only recently been fully realized. Biodeterioration may be defined as any undesirable change in the properties of a material caused by the vital activities of organisms, and while a detailed account of microbial processes without direct bearing on metals would be out of place in this book, it is important to remember that situations favourable to the corrosion of metals can be induced by microbes acting on protective coatings or on the fluids contained in metal pipes and vessels. Fig. 1.11 represents a synthesis of the principal methods of microbial corrosion and the principal microbial activities that provide opportunities for the metal corroders to start work.

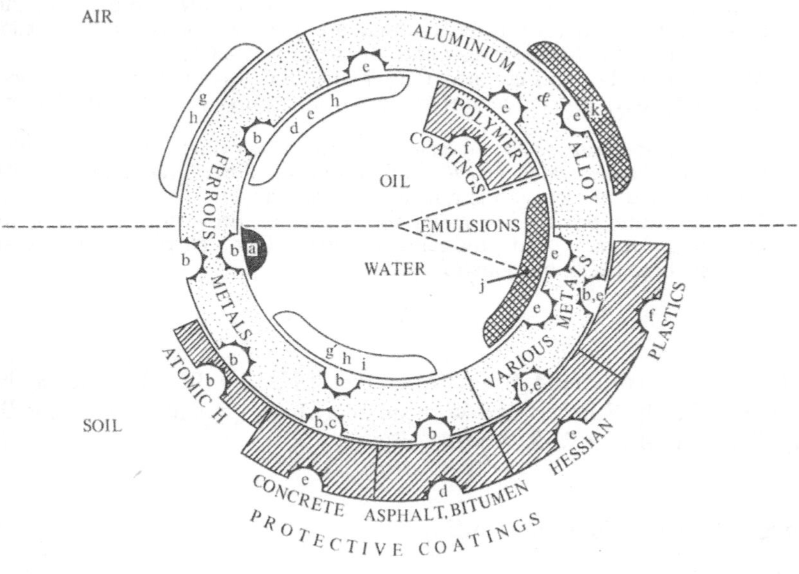

Fig. 1.11 *Diagram to illustrate the principal methods of microbial corrosion of metals and protective materials*

KEY

a Tubercle (inorganic + iron bacteria) causing corrosion by differential aeration and providing environment for *b*.

b Anaerobic sulphate-reducing bacteria, able to remove protective hydrogen film on metal under oxygen-free conditions, in ground or under water, and cause sulphide corrosion of the metal.

c Sulphur-oxidizing bacteria. Produce sulphates and free sulphuric acid; break down concrete, attack metals (corrode buried metal pipes, concrete sewers; also responsible for acid mine waters which corrode pumping machinery etc.).

d Hydrocarbon utilizers. Break down aliphatic and bituminous coatings, allowing access of *b* to metal beneath; grow in fuel oil tanks etc. if traces of water present, forming slime and probably acids, and providing nutrients for other microbes.

e Various microbes producing organic acids as end-products of growth, attacking mainly non-ferrous metals and certain coatings.

f Bacteria and moulds able to break down polymers, *e.g.*, synthetic rubber and polyurethane coatings in fuel tanks, and having a possible role in corrosion; or able to embrittle vinyl polymer coatings by degrading the plasticizers.

g Mainly algae, forming slimes on above-ground damp surfaces, and constituting their main hazard when falling or being carried to position *g′* under water where they produce the same effect as *h*.

h Slime-forming moulds and bacteria (some of which may produce organic acids, or utilize hydrocarbons), which provide conditions for the setting up of differential aeration cells and for the growth of *b*.

i Mud on river bottoms etc. providing matrix for heavy growth of microbes (including oxygen-free conditions for *b*) and causing corrosion of piers, piles etc.

j Sludge (inorganic debris – scale, corrosion products etc. – providing matrix for concentrated growth and setting up differential aeration cells; and organic debris providing nutrients for growth).

k Debris (mainly organic) on metal above ground, providing conditions for growth of organic acid-producing microbes (*e.g.*, corrosion of aircraft bodies where traces of foodstuffs collect).

* * *

Fig. 1.1 is reproduced from "Microbial Life" by W. R. Sistrom, by courtesy of Holt, Rinehart and Winston Inc.; Figs. 1.3 and 1.4 were kindly supplied by Professor R. S. Wolfe.

Suggestions for Further Reading

General Microbiology including Cultivation

POSTGATE, J. R.: *Microbes and Man*, Penguin Books Ltd., Harmondsworth, Middlesex, 1969.

SISTROM, W. R.: *Microbial Life*, Holt, Rinehart and Winston Inc., New York and London, 1962.

STANIER, R. Y., DOUDOROFF, M. AND ADELBERG, E. A.: *General Microbiology*, 2nd Edn., Macmillan and Co. Ltd., London, 1963.

Microbial Chemistry

ROSE, A. H.: *Chemical Microbiology*, 2nd Edn., Butterworth and Co. Ltd. London, 1968.

Industrial Microbiology

ROSE, A. H.: *Industrial Microbiology*, Butterworth and Co. Ltd., London, 1961.

DAVIS, J. B.: *Petroleum Microbiology*, Elsevier Publishing Co. Ltd., London, 1967.

Sterilization

SYKES, G.: *Disinfection and Sterilization*, 2nd Edn., E. and F. N. Spon Ltd., London, 1965.

RICHARDS, J. W.: *Introduction to Industrial Sterilization*, Academic Press Ltd., London, 1968.

Biodeterioration

WALTERS, A. H. AND ELPHICK, J. J. (Editors): *Biodeterioration of Materials*, Elsevier Publishing Co. Ltd., London, 1968.

CHAPTER TWO: I. A. MENZIES

Introductory Corrosion

Introduction

The subject of corrosion is a very broad interdisciplinary study involving many aspects of chemistry, metallurgy and chemical engineering. In recent years the subject of microbiological corrosion has begun to develop and to be recognized, and this area of corrosion is of enormous significance in relation to the behaviour of materials buried in soil, in industry in general in relation to all types of water usage, in the machining and fabrication of materials and in certain advanced technology areas such as aircraft. As new materials are developed we shall have to cope with biodeterioration of these and since these may in some cases take the form of coatings on metals the implication of this with respect to conventional corrosion processes will have to be taken into account. Indeed there is now a very strong case for considering corrosion as the "environmental degradation of materials" and for allowing this term to embrace all materials and all forms of degradation irrespective of mechanism.

This article will be concerned only with those aspects of corrosion which are immediately relevant to the scope of this book. Atmospheric corrosion, mechanical influences in corrosion, *i.e.*, stress corrosion, corrosion fatigue, cavitation and fretting, and considerations of the effects of temperature and flow on corrosion will not be covered. The reader who may wish to extend his knowledge of these aspects of corrosion is advised to consult the standard texts available. Electrochemistry is excellently treated in a new major work[1] whilst a more concise account[2] will be helpful to the beginner who requires an up to date but much less detailed treatment of the subject. Other major texts deal with most aspects of corrosion at some length[3–8] whilst the beginner may first wish to consult some of the introductory monographs[9–11]. Various aspects of corrosion including the fundamental principles, test methods, selection of materials and corrosion control have been reviewed in a previous monograph[12].

In order to appreciate the immediate relevance of the more important aspects of corrosion which will be discussed here it should be noted at the outset that corrosion brought about by microbial action usually occurs by a number of mechanisms which are, in themselves, reasonably well understood as corrosion phenomena *per se*. Some of the major mechanisms of interest here in relation to microbial action are:

1. the production of corrosive metabolic products;
2. the production of differential aeration and concentration cells;
3. depolarization of cathodic processes;
4. disruption of natural and other protective films; and
5. the breakdown of corrosion inhibitors.

Since we have discovered in Chapter 1 the basic microbiological processes which occur it is now essential to provide the basic background to corrosion processes and in particular to discuss the background to the above mechanisms in order that microbiological aspects of corrosion can be discussed in more detail in a variety of circumstances in Chapters 3 to 6.

Classification of Corrosion Processes

The major divisions of corrosion processes are summarized in Fig. 2.1. We can see here that there are two broad divisions of corrosion as between electrolytic processes and metal-gas reactions which include atmospheric corrosion and direct reactions between metals and gases at all temperatures. These metal-gas reactions are film-forming processes and atmospheric corrosion involves film formation and film breakdown processes. We shall, however, concern ourselves with electrolytic corrosion processes in this Chapter. These processes can be divided into two main groups in which corrosion takes place in electrolyte solutions and in which corrosion occurs under conditions where metals are buried in the soil. Both of these groups are important in relation to microbiological aspects of corrosion and various aspects are dealt with in the following Chapters. In these groups further division can be made into

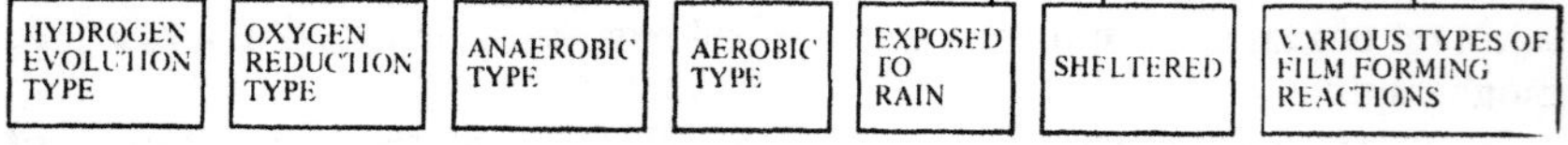

Fig. 2.1 *Simplified classification of corrosion processes*

conditions where oxygen is absent, *i.e.*, hydrogen evolution reactions under immersed conditions and anaerobic conditions underground, and conditions where oxygen is present and plays a significant role, *i.e.*, where oxygen reduction takes place under immersed conditions and underground corrosion under aerobic conditions. In order to deal with these matters further it is necessary to introduce basic concepts of electrolytic cells and some of the terminology relating to corrosion processes.

The Electrode Potentials of Metals, Electrolytic Cells, Anodic and Cathodic Processes, and the Principles of Control

Any individual metal establishes a potential with respect to a solution though this potential cannot be measured in absolute terms. We can however measure the potential difference between it and another electrode. If we choose an appropriate second electrode which has a constant potential, *i.e.* a reference electrode, we can then investigate changes in the potential difference and relate these to the metal electrode under investigation. We can choose the hydrogen electrode as the reference electrode and if we choose conditions such that the hydrogen ion activity (*i.e.* the thermodynamic hydrogen ion concentration) is unity and the hydrogen pressure is 1 at. we have the standard hydrogen electrode. The precise values for the electrode potentials of metals *under clearly defined conditions* can then be tabulated. In electrochemistry the *standard electrode potential* of a metal is the e.m.f. of the cell

$$M \,|\underbrace{M^{z+},\ H^{+}}_{\text{unit activity}}|\, H_2,\ Pt$$

at 25°C. Again the metal ion activity is always unity. Under these conditions the e.m.f. of the hydrogen electrode is, by convention, taken to be zero. We can thus derive the electrochemical series of the elements (Table 2.1). In corrosion processes we can obtain some useful information by considering the electrochemical series, although *in practice* its usefulness is much less than we might expect. This stems, in part, from the fact that, in practice, the actual conditions in solution are often far removed from the standard conditions defined above, and this is especially the case where complexing ions may be present. Under these conditions the order of two particular metals may even be reversed.

Table 2.1. *Standard electrode potentials of metals in aqueous solutions on the hydrogen scale* (25°*C.* 1 *atm.*)

	Metal/Metal ion	Electrode potential (Volts)	Metal/Metal ion	Electrode potential (Volts)	
Base	Li/Li^{+}	−3·04	Co/Co^{2+}	−0·28	
	K/K^{+}	−2·93	Sn/Sn^{2+}	−0·14	
	Ba/Ba^{2+}	−2·90	Cu/Cu^{2+}	+0·34	
	Ca/Ca^{2+}	−2·87	H_2/H^{+}	0	
	Na/Na^{+}	−2·71	Ni/Ni^{2+}	−0·25	
	Mg/Mg^{2+}	−2·37	Pb/Pb^{2+}	−0·13	
	Al/Al^{3+}	−1·66	Cu/Cu^{+}	+0·52	
	Ti/Ti^{2+}	−1·63	Hg/Hg^{2+}	+0·85	
	Zn/Zn^{2+}	−0·76	Ag/Ag^{+}	+0·79	
	Fe/Fe^{2+}	−0·44	Hg/Hg^{2+}	+0·78	
	Cd/Cd^{2+}	−0·40	Au/Au^{3+}	+1·5	Noble

As written any element will displace one below it in the series from a solution of its ions, e.g. Fe will liberate Cu from a $CuSo_4$ solution.

The Electrochemical Character of Corrosion and Basic Nomenclature
A simple *voltaic* cell may be regarded as a model corrosion cell in which two dissimilar metals are coupled together in a conducting electrolyte. This type of system will, however, merely *illustrate* the electrochemical nature of corrosion and basic nomenclature (cf. Fig. 2.2). Here we can see that two different metals will each have different potentials in the electrolyte. The potential *difference* between them results in the setting up of the electric current and corrosion of the more reactive element (anode) in the couple. The magnitude of the corrosion that takes place is quantitatively related to the magnitude of the current. At the anode positively charged metal ions enter the electrolyte leaving excess electrons on the metal surface which becomes negatively charged. These electrons then flow through the external circuit to the

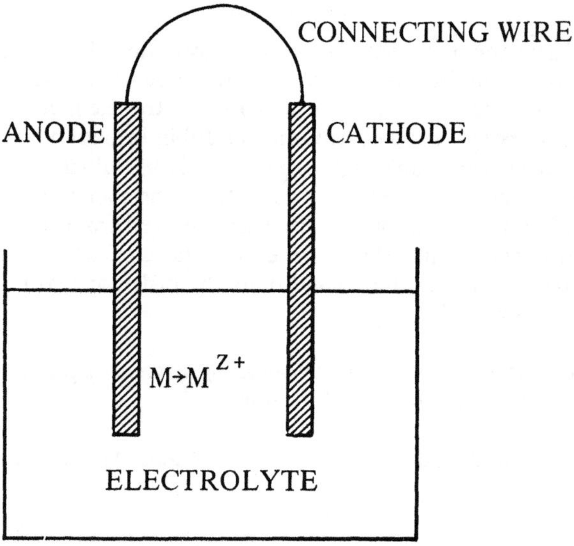

Fig. 2.2 *Simple voltaic cell*

cathode, *i.e.* the less reactive member of the couple, which then receives protection. These general features find their counterpart in ordinary immersed corrosion in aqueous solutions or in natural waters and are not confined or restricted to contacts between dissimilar metals. On the surface of an immersed single metal various factors bring about the presence of anodic and cathodic areas, each of which makes its characteristic contribution to the corrosion process.

Although the electrode potential relationships of the anode and cathode determine the magnitude of the unpolarized e.m.f. of the system this becomes modified as soon as current begins to flow with a corresponding reduction in the magnitude of the current density on which the magnitude of corrosion depends. At the anodes, films, which tend to decrease the ease of emergence of metal cations, may form and may slow down the corrosion rate. This is known as *anodic polarization.* Similarly, if incoming electrons from the anode cannot be released easily and tend to accumulate at the cathode the current density in the cell is reduced and hence corrosion of the anode is reduced. In this case the system is *cathodically polarized.* In fact it can be seen that a corrosion process can be controlled by controlling the situation at either the anode or the cathode.

There are two main ways in which the electrons which arrive at cathodes can be removed and these give rise to the two major divisions of immersed corrosion.

Type I – Hydrogen Evolution

This reaction usually takes place in acid solutions in the absence of oxygen and may be represented by

$$2H^{+}_{\text{(aq.)}} + 2e_{\text{(metal)}} \rightarrow H_{2\,\text{(gas)}} \qquad (1)$$

The amount of hydrogen discharged must correspond exactly to an equivalent amount of metal dissolved by corrosion at the anode. In this process cathodic polarization may arise from (*a*) depletion of hydrogen ions in the immediate vicinity of the cathode (concentration overpotential) or (*b*) hydrogen overpotential which is a measure of the difficulty of hydrogen evolution (activation overpotential). In considering the overall reaction depicted in equation (1) there are a number of detailed steps which have to be considered.

1. *Transfer of hydrated proton from bulk of solution to the electrode surface*

$$H_3O^{+}_{\text{(bulk)}} \xrightarrow[\text{convection}]{\text{migration, diffusion}} H_3O^{+}_{\text{(electrode)}}$$

A necessary part of the discharge step is the adsorption of the resulting hydrogen atom on the electrode surface.

2. *Discharge step*

$$H_3O^{+}_{\text{(electrode)}} + e_{\text{(from metal M)}} \xrightarrow[\text{discharge}]{\text{dehydration}} \underset{\substack{\text{hydrogen atom}\\ \text{adsorbed on M.}}}{M\text{–}H} + H_2O$$

3. *Combination step*

$$M\text{–}H + M\text{–}H \xrightarrow[\text{combination}]{} H_{2\,\text{(gas)}}$$

An alternative step instead of the above combination reaction is as follows:

4. *Secondary discharge*

$$H_3O^{+}_{\text{(electrode)}} + M\text{–}H + e \xrightarrow[\text{discharge}]{\text{secondary}} H_{2\,\text{(gas)}}$$

Overpotential associated with step (1) is known as *concentration overpotential.* That associated with steps (2), (3) and (4) which are rate determining is known as *activation overpotential.*

In the case of hydrogen evolution, activation overpotential is the most fundamental issue. For slow chemical reactions the activation energy is usually found by measuring the rate constant of the process at different temperatures and applying the Arrhenius equation

$$k = \text{const.} \times e^{-E_A/RT} \tag{2}$$

where E_A is the activation energy. For electrochemical processes the rate of reaction is measured by the current passing at constant temperature and we can follow how this quantity varies as the potential applied to the electrode is varied over a wide range, *i.e.* we can determine the polarization curve. For overpotentials (η) > 0·05 V we can relate current density (i) to overpotential by the Tafel equation

$$\eta = a + b \log i \tag{3}$$

Experimentally determined values of "a" and "b" are characteristic for each electrode process. For hydrogen evolution a "b" value of 0·03 is characteristic of a rate-determining combination step whereas a "b" value of 0·12 is characteristic of a rate determining discharge or secondary discharge step. Thus hydrogen overpotential is important and may well control the cathodic process and hence the rate of corrosion.

Type II – Oxygen Reduction

In natural waters including neutral salt solutions and for most of the industrially important ferrous materials, accumulation of electrons at the cathodes is prevented mainly by intervention of oxygen. We may regard oxygen reduction as represented by

$$O_2 + 4e + 2H_2O \rightarrow 4OH^- \tag{4}$$

with the production of hydroxyl ions near the cathode surface and this probably represents the situation in normal immersed corrosion in neutral solutions. Alternatively we can regard the situation as represented by

$$4(M\text{–}H) + O_2 \rightarrow 2H_2O + 4\,M \text{ (in lattice)} \tag{5}$$

which we can consider to be an alternative means of removing hydrogen atoms produced by the discharge step (2) above. This step is energetically easier than any of the desorption steps and hence, by making the surface area available to hydrogen greater, it has the effect of increasing the rate of the discharge step. Hence on adding an aqueous solution of oxygen to an irreversible hydrogen electrode system an increased current passes and the overpotential falls or disappears completely. The electrode is *depolarized* and oxygen is the *depolarizer.* In the majority of cases of corrosion by oxygen reduction, diffusion of oxygen to the metal surface rather than activation will be rate controlling. Thus in a corrosion process where the cathodic reaction is oxygen reduction we can control the corrosion rate by controlling the rate of supply of oxygen to the system.

Having presented the general ideas concerning corrosion processes and indicated the main divisions, terminology and processes it is now possible to deal with the subject in greater detail. Finally we can examine the thermodynamic approach to corrosion and having done this and realized the limitations it is possible to consider the kinetic aspects of electrode processes and to see that these are perhaps of even greater significance.

Thermodynamic Aspects of Corrosion

The various possible anodic reactions of a metal in an aqueous electrolyte, in the absence of complexing or precipitating ions other than the hydroxyl ion, follow one of the paths below[13]

$$M \rightleftharpoons M^{x+}{}_{(aq.)} + xe \tag{6}$$

$$M + xH_2O \rightleftharpoons M(OH)_{x\ (solid)} + xe + xH^+ \tag{7a}$$

$$M + xOH^- \rightleftharpoons M(OH)_{x\ (solid)} + xe \tag{7b}$$

$$M + xH_2O \rightleftharpoons M_x{}^{x-}{}_{(aq.)} + 2xH^+ + xe \tag{8a}$$

$$M + xOH^- \rightleftharpoons MO_x{}^{x-}{}_{(aq.)} + xH^+ + xe \tag{8b}$$

or their equivalents in terms of hydroxy-oxides and metal oxyanions. The solid product $M(OH)_x$ can be considered to be typical of a sparingly soluble solid hydroxide which might have attached water of hydration and there may possibly be formation of monomeric or polymeric forms of oxides or hydroxides. The species $M^{x+}{}_{(aq.)}$ and $MO_x{}^{x-}{}_{(aq.)}$ are normal water-soluble cations and oxyanions respectively. The pairs of equations 7a and b, and 8a and b, represent reactions which are thermodynamically identical except for the choice of standard states. All of the reaction schemes are overall processes and intermediate mechanistic steps are not presented.

The electrode potentials corresponding to the half cells having as their reactions the above processes (6) to (8) have either been measured directly or computed for many metal-H_2O systems. These electrode potentials have been combined[14] with data for the solubilities of oxides and hydroxides and with equilibrium constants for the reactions to produce potential/pH equilibrium diagrams known as Pourbaix diagrams at 25 °C (cf. Fig. 2.3). In these diagrams the variations of the half-cell potentials and of the solubilities of the solid phases with pH are plotted for a series of values of metal ion activity. Corrosion has to be defined in terms of an equilibrium activity of metallic anions or cations in solution and an arbitrary criterion for corrosion of 10^{-6} g ion/l has been adopted by Pourbaix. In discussing the corrosion of a metal in general terms a simplified form of the Pourbaix diagram (Fig. 2.4) is

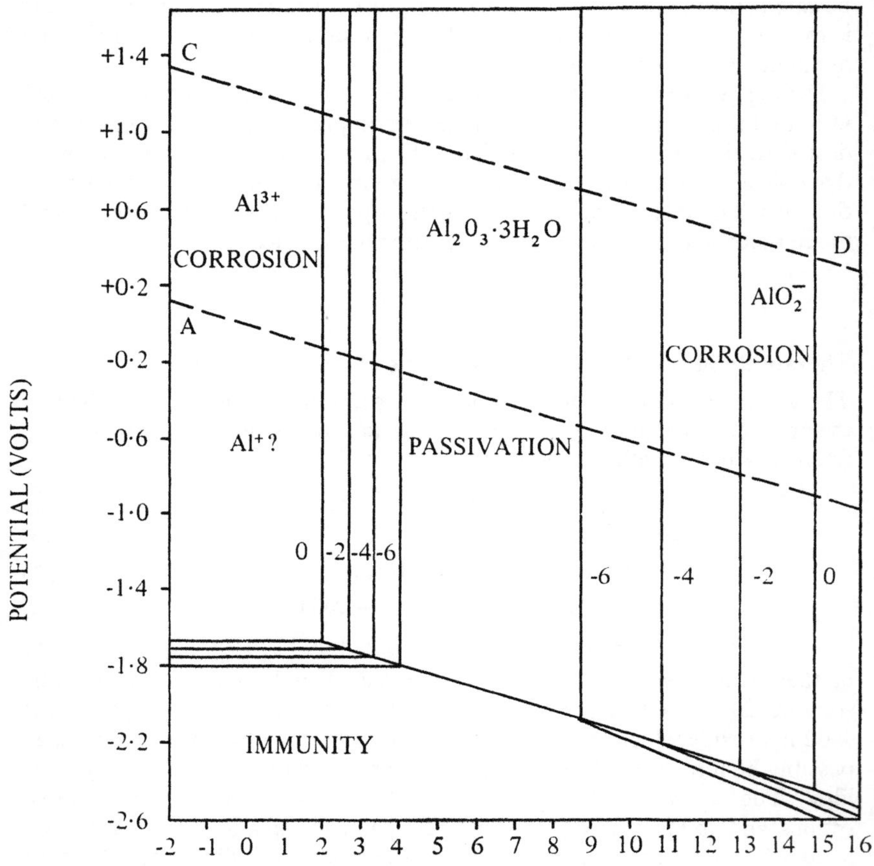

Fig. 2.3 *Potential/pH diagram for Al/H_2O system at 25°C*

often used. The arrows (6)–(8) in Fig. 2.4 indicate the anodic reactions represented in equations (6)–(8) above. Hence at low values and high values of pH corrosion should occur forming Al^{3+} in acid solutions and AlO_2^- in alkaline solutions. At intermediate values of pH, $Al_2O_3.3H_2O$ is formed according to equation (7) and aluminium should be passive in this region. The effectiveness of the passive film of solid product in reducing or preventing corrosion depends upon the nature of the film, its adhesion, and its mechanical properties. It should be noted that in the zone labelled "immunity" corrosion is not possible on energetic grounds whereas in the zone labelled "passivation" corrosion should not occur provided that the film is formed in such a way that it has the necessary physical, structural and mechanical properties[5]. One further point should be made concerning the reactions in equations (7) and (8). The pH in the solution near the anode will fall because of the production of hydrogen ions or removal of hydroxyl ions[15–17] and hence if the solution is not buffered we may move from the passive zone to the active zone (Fig. 2.4). In order that anodic dissolution of metals like Fe, Ni, Sn, Zn and Al may take place as soluble metal ions in neutral solutions it is necessary for initial production of anodic acidity to occur[15–18]. If the solution is buffered then passivation may be immediate[19–21]. However, most likely buffering agents either form salts which can be co-precipitated or form easily soluble complexes.

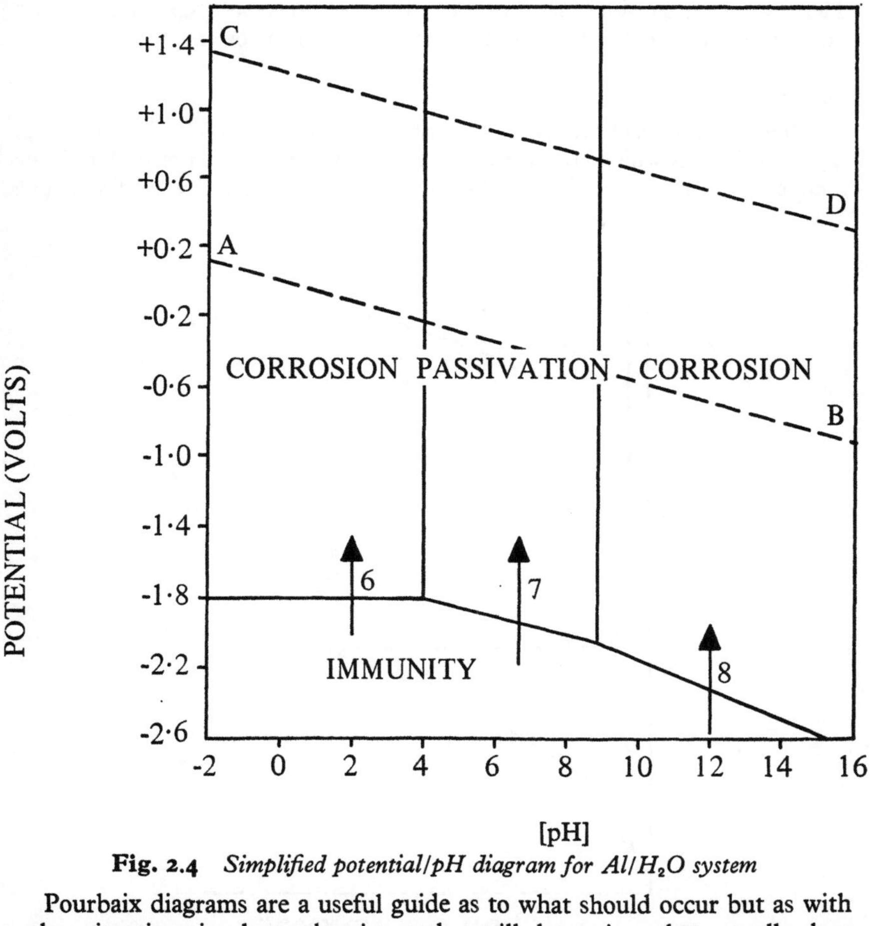

Fig. 2.4 *Simplified potential/pH diagram for Al/H_2O system*

Pourbaix diagrams are a useful guide as to what should occur but as with other situations in electrochemistry what will determine what actually does occur in practice are the rates of the processes, and it is necessary to combine thermodynamic data with kinetic data and a knowledge of the properties of the passivating film in order to gain a complete understanding of corrosion phenomena. Furthermore, in most real situations, complexing ions are present to some extent and acids are often aerated. Films may not always be precipitated in the most appropriate form to provide the protection required and hence great care must be taken in applying Pourbaix diagrams in practice and agreement between theory and practice should certainly *not* be expected on all occasions.

Electrode Kinetics and Polarization Characteristics

Here we can consider anodic and cathodic processes separately before combining them to examine the situation in natural corrosion.

Anodic Processes

A schematic polarization curve, obtained potentiostatically, is shown in Fig. 2.5. In solutions which are sufficiently acid (low pH), or are not buffered, the anodic current and hence the dissolution rate will increase as the anode

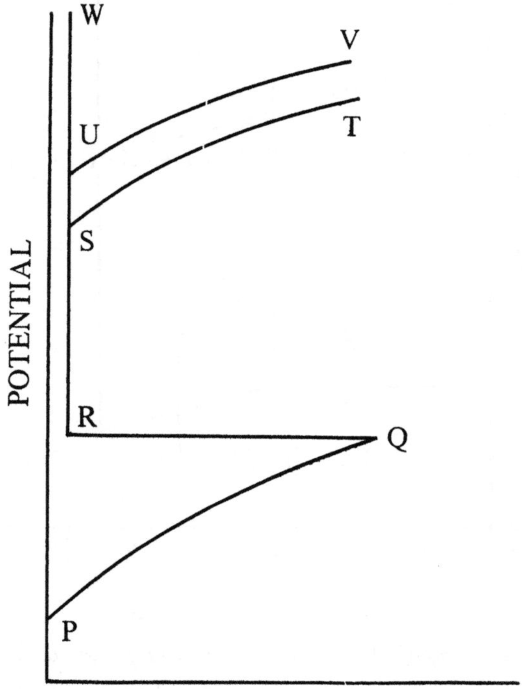

Fig. 2.5 *Schematic anodic polarization curve*

potential is made more positive (P–Q, Fig. 2.5). When the potential becomes more positive then formation of oxide or hydroxide is thermodynamically possible. If this is also favoured kinetically the formation of a film will act as a barrier to further dissolution and the current will fall, Q–R in Fig. 2.5, and passivation of the metal occurs[22]. Further rises in anode potential along R–S do not cause significant increase in current density until very much higher potentials are attained[23]. Here two possibilities which depend upon the nature of the film occur. If the film is a good electronic conductor it may be oxidized to a soluble species and dissolution recommences along S–T. In this region the metal is in the transpassive state. This situation is of considerable practical interest when stainless steels are used in strongly oxidizing conditions where oxidation of Cr(III) to Cr(VI) may occur and where CrO_3 is readily soluble. Alternatively, if the film is a good electronic conductor oxygen evolution may occur along U–V. If, on the other hand, the film is a poor electronic conductor then under conditions of high anodic potential high field strengths arise and anodic oxidation or anodizing may take place along U–W. This is operated commercially as a process for the protection of aluminium and anodizing of this metal is relatively well understood. In natural corrosion processes the anode potential cannot normally be polarized such as to raise the potential above S, one exception to this being stainless steels in strong nitric acid. It should be noted that the presence of halide ions, and especially chloride ions, may prevent or delay passivation and that passive films will tend to break down in the presence of such ions. If, however, the metal chloride is of limited solubility, then a less effective form of passivity may arise due to the formation of chloride films. Molybdenum and silver are two examples of metals of this type and commercial alloys containing ~ 30% Mo, 6% Fe and 64% Ni have been developed[24] for plants handling hot hydrochloric acid under non-oxidizing conditions.

Cathodic Processes

As we have already seen the two most important cathodic processes are the reactions in equations (1) and (4). The lines AB and CD in Figs. 2.3 and 2.4 represent the potentials of the reactions

$$2H^+ + 2e \rightleftharpoons H_2 \text{ and}$$
$$2OH^- \rightleftharpoons \tfrac{1}{2}O_2 + H_2O + 2e$$

i.e. the hydrogen and oxygen evolution lines. Below the hydrogen line hydrogen gas is liberated whilst above the oxygen line oxygen is evolved. Between the lines in the diagram both are stable, *i.e.* water is stable. Alternative cathodic processes may involve the reduction of ions such as Fe^{3+} or MnO_4^- or molecular species such as nitric acid. For cathodic reactions, polarization in the negative direction increases the current density and a number of examples are shown in Fig. 2.6. It can be seen that reduction of

HNO_3, Fe^{3+} and H^+ are relatively easy processes whereas the reduction of oxygen requires a relatively high polarization for a small current density and a limiting current density arises at a relatively low current density value due to the limited solubility of oxygen in aqueous solutions. Similar limiting current densities arise for the other species but at much higher values owing to the greater solubilities of the other species.

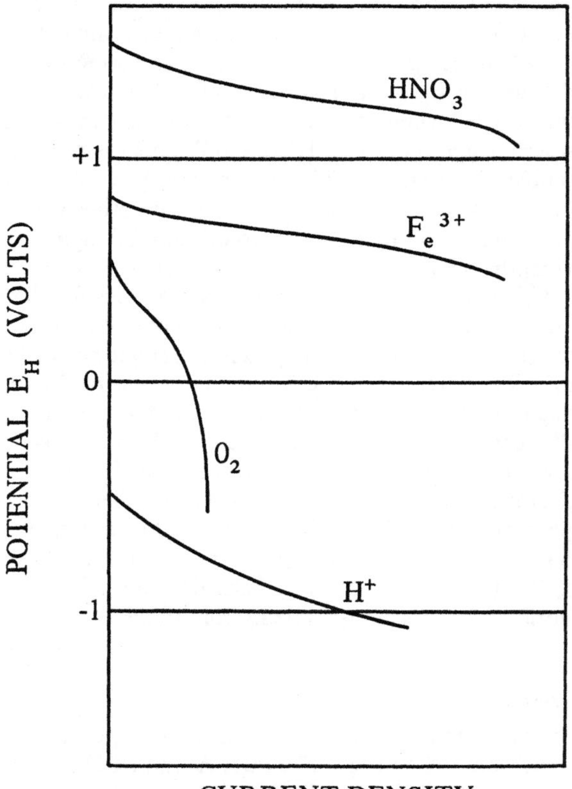

Fig. 2.6 *Schematic cathodic polarization curves (after Hoar)*

Natural Corrosion

When natural corrosion takes place on a single metal the anodic and cathodic reactions mutually polarize each other. This can be seen in relation to an Evans[25] diagram (Fig. 2.7) which relates the potential (E) and current (I). When no current is flowing ($I = 0$), E_a and E_c represent the reversible equilibrium potentials of the corresponding half reactions of the corrosion process. In the dissolution of a metal M in normal acid solutions ($a_{H_3O^+} = 1$), E_a would represent the potential of the reversible metal electrode $M|M^{z+}$ at

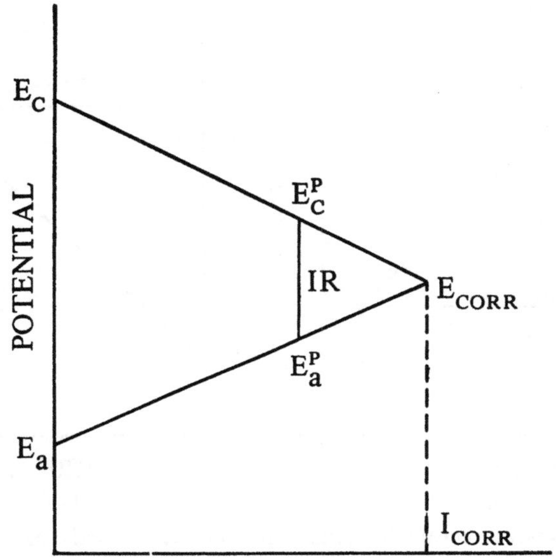

Fig. 2.7 *Evans potential/current diagram*

$a_m{}^{z+} = 10^{-6}$ g. ion/l, and E_c would be the potential of the hydrogen electrode in the same solution. If the reaction is to occur spontaneously E_c must be more noble than E_a. When mutual polarization occurs the anodes become more noble and the cathodes more base. Polarization to the values $E_a{}^P$ and $E_b{}^P$ occurs. The working e.m.f. $(E_c{}^P - E_a{}^P) = IR$ where I is the current flowing through the circuit and R is the total resistance due to electrolyte, films, etc. The value of the corrosion potential measured under these circumstances will vary between $E_c{}^P$ and $E_a{}^P$ depending upon the position of the probe electrode. When the conductivity of the electrolyte is sufficiently high R is sufficiently small for $E_a{}^P \simeq E_c{}^P$ to within 1 mV and each may be identified as the corrosion potential E_{CORR}. This represents the steady state or mixed potential of the system and the corrosion current is I_{CORR}. This is a special case of mixed potentials which arise when two or more electrode processes are simultaneously in operation on a metal surface and there is no ohmic (IR) drop between them[26–28].

The diagram in Fig. 2.7 applies when a system is under both anodic and cathodic control, *i.e.* when both the anodic and cathodic electrode processes affect the situation, and this is known as mixed control. In some extreme cases where the electrolyte resistance or overall resistance is very high we have the situation as depicted with $E_c{}^P$ and $E_a{}^P$ and this is known as ohmic control. In other situations we may have cathodic control (Fig. 2.8a) or anodic control as in Fig. 2.8b. Real polarization curves are not usually straight lines as shown in Figs. 2.7 and 2.8 and detailed theoretical analyses of the shapes of

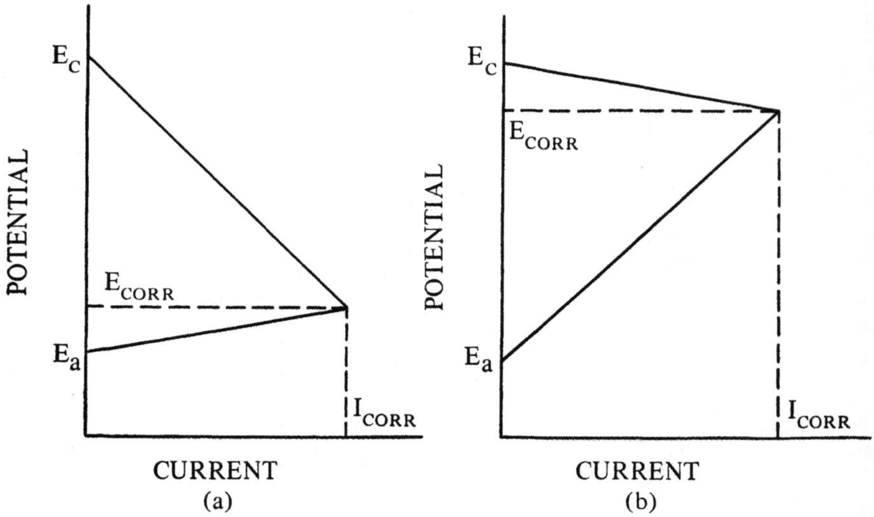

Fig. 2.8 *Evans-type current/potential diagrams: (a) cathodic control, (b) anodic control*

polarization curves have been made[29] and methods of calculating corrosion rates from polarization data have been discussed[30,31]. Some typical curves are shown in schematic form in Fig. 2.9. It is clear (Fig. 2.9a) that the potentials

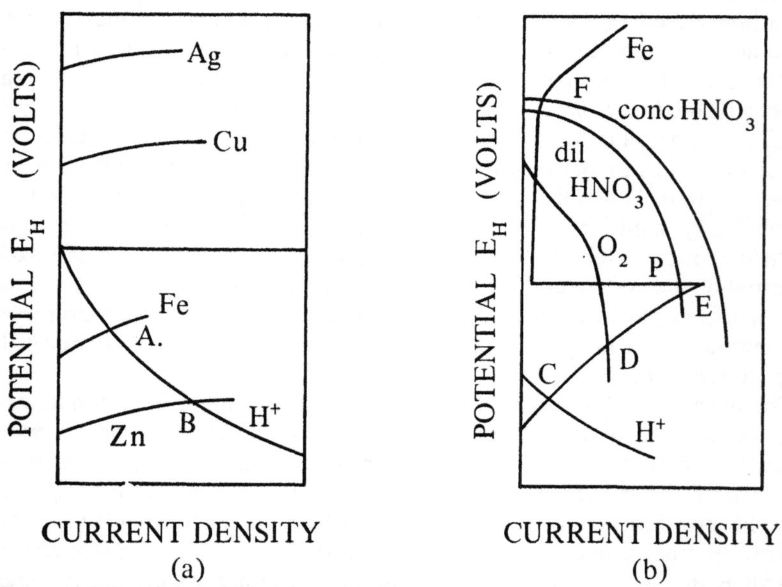

Fig. 2.9 *Schematic polarization curves showing mixed potentials (after Hoar): (a) de-oxygenated acid solutions, (b) oxygenated acid solutions*

required for the anodic dissolution of silver and copper are greater than can be attained by hydrogen ion reduction, the only possible cathodic process in de-oxygenated acids. Iron dissolves at the rate and potential corresponding to point A whereas the more active metal B dissolves at the faster rate represented by point B. In this situation, however, impurities in the zinc may have an appreciable effect upon the corrosion rate actually observed. If impurities with high hydrogen overpotential (*e.g.* Pb, Hg) are present the corrosion rate of zinc will be decreased whereas if impurities with low hydrogen overpotentials are present (*e.g.* Cu, Fe) the corrosion rate will be increased[32]. If we now consider Fig. 2.9b iron will dissolve at the rate and potential corresponding to point C in the absence of oxygen. However, in the presence of oxygen, the potential of iron is raised to that at point D and there is a corresponding increase in the rate of corrosion. If dilute HNO_3 which functions as a mild oxidizing agent is present there is a further increase in the potential of iron to point E and a further increase in corrosion rate. If a much stronger oxidizing agent is present (*e.g.* concentrated HNO_3) the potential of iron is raised above the passivating potential P and a passivating film is formed. This reduces the corrosion current to F with a corresponding rise in potential. Although such passive films can remain intact in, say, oxygenated dilute HNO_3 for considerable periods of time, they will not be produced under these conditions.

Anodic and Cathodic Protection and the use of Inhibitors

It is possible to hold the potential of a metal in the passive range (R–S in Fig. 2.5) by means of an external cathode and a potentiostatic controller[33] and this is known as anodic protection[34]. The technique has been examined both in the laboratory and in the field[35–47]. The use of the technique is somewhat limited in practice since the metallic structure must be of a material of suitable composition for passivation in the particular solution which is in the plant. If protection breaks down at ANY point corrosion will be extremely rapid at that point because of the low resistance path formed. This means that extensive monitoring and control facilities are required. Furthermore, the passive films will be destroyed if aggressive ions such as chloride ion are present. One exception to this is titanium which can perform satisfactorily even in hydrochloric acid solutions[42,43]. Although throwing power in such systems is not always good prior to the establishment of passivation it is very much better once passivation has taken place[34]. It has been used in the fertilizer production industry to control corrosion of mild steel in contact with ammonia–ammonium nitrate solutions[35]. The technique has many potential applications once individual monitoring problems have been overcome.

Cathodic protection has been more widely used[48,49]. It can be shown that any action which is taken and which results in the lowering of the potential of a dissolving anode decreases the anodic current density and therefore the rate of corrosion. If, for example, the corroding metal is connected to an external anode, *i.e.* a metal which is baser or more active than the corroding metal, or a material with its potential so adjusted by application of an external e.m.f., there is a decrease in anodic current density and corrosion (cf. Fig. 2.10). If the potential of the metal is reduced from E_{CORR} to E'_{CORR} then the external current is represented by B − E'_{CORR} and the corrosion current is reduced from I_{CORR} to I'_{CORR}. It follows that if the potential is reduced to E_a then the corrosion current will be zero and the external current will be represented by $E_aE''_{CORR}$. Under these conditions there will be complete cathodic protection[50,51]. There is usually an increase in the alkalinity of the environment produced by the cathodic reaction and care has to be taken to ensure that this does not become excessive since many metals (*e.g.* Fe, Al, Zn) form soluble oxyanions under high pH conditions. If paints are used in conjunction with cathodic protection these must also be able to withstand the alkalinity of the medium. This form of protection is normally applied to ferrous metals buried in the earth or immersed in water, often sea-water. Thus it is usual to protect underground pipelines, foundations of various types, certain harbour installations, and ship's hulls. Magnesium alloys **(Mg—6%Al—3%Zn—0·2%Mn)** and aluminium alloys **(Al—6%Zn)** are often used as sacrificial anodes, the alloying elements usually being present to prevent film formation at the anode. Impressed current systems using inert anodes such as carbon, platinum and stainless steel are often used since closer

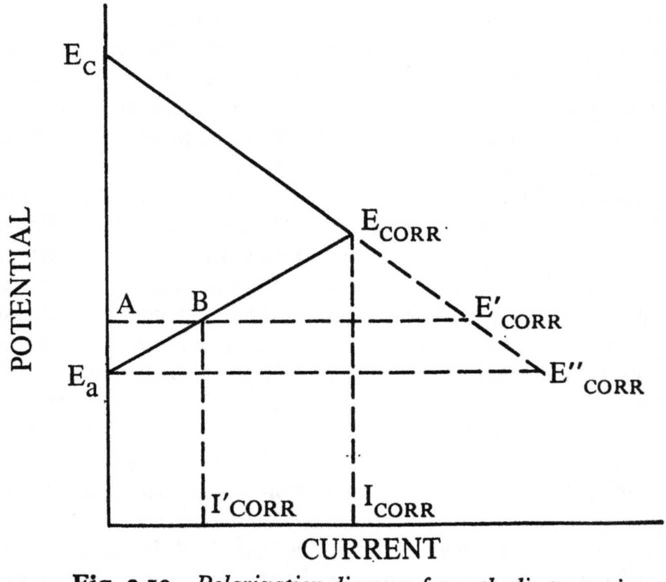

Fig. 2.10 *Polarization diagram for cathodic protection*

control of the system is possible and it has economic advantages. Cathodic protection of pipelines can be complicated by variations of oxygen permeability in different soils along the length of the line leading to the setting up of differential aeration cells (see next section) and by bacterial corrosion (see Chapter 3). In practice, cathodic protection is usually used in conjunction with bitumen, fibre glass, plastics or paints which reduce the current requirements.

Anodic and Cathodic Inhibition

Anodes which are active may be inhibited by adsorption of organic molecules, or by plugging with precipitated material or by other means. Anodic inhibitors are of wide variety and include chromates and nitrites which are oxidizing anions and other compounds such as silicates, phosphates, benzoates, molybdates and tungstates which are non-oxidizing. All of these function in neutral or alkaline solutions. In deaerated solutions oxidizing inhibitors lead to the formation of γFe_2O_3:

$$Fe + Na_2CrO_4 + H_2O \rightarrow \gamma Fe_2O_3 + Cr_2O_3 + NaOH$$

which is highly protective and in the case above Cr_2O_3 is also incorporated into the growing film[52]. The anodic reaction becomes highly polarized and the

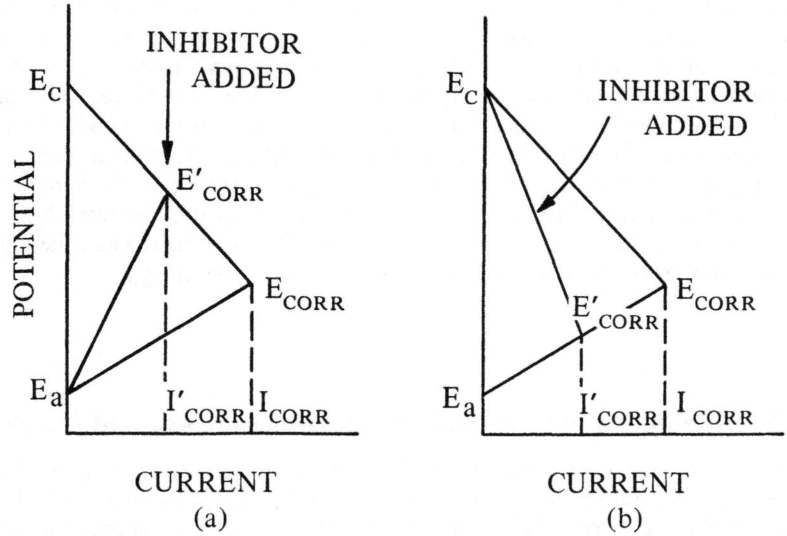

Fig. 2.11 *Schematic polarization curves illustrating inhibitor behaviour (after Hoar): (a) anodic inhibitor added, (b) cathodic inhibitor added*

mixed corrosion potential of the specimen under such conditions is shifted in the noble direction as shown in Fig. 2.11a and the current falls[53,54]. If non-oxidizing inhibitors are used oxygen must be present in the environment. If oxygen is present when oxidizing inhibitors are used the minimum effective inhibitor concentration is reduced. This is the opposite role of oxygen to that previously discussed where it functioned as a cathodic depolarizer. In acidic solutions, γFe_2O_3 will not form and oxygen functions only as a cathodic depolarizer. From this it can be seen that both pH and oxygen content are important in determining the effect of anions in inhibiting the corrosion of iron.

If access of the cathodic reactant to the metal surface is hindered both the cathodic process and corrosion will be slowed down. This can be implemented by a reduction in the oxygen content of the environment or by non-conducting films as may occur in hard water. In this case the mixed potential is lowered and there is again a decrease in corrosion current (Fig. 2.11b). The oxygen content of boilers may be reduced by hydrazine additions[55,56]. The carbon dioxide content of water is important and it may be present as carbonate, bicarbonate, dissolved CO_2 which maintains the carbonate/bicarbonate mixture equilibrium or free aggressive carbon dioxide. Hard waters contain calcium and magnesium salts which actively participate in scale formation and such waters have to be treated to allow only thin inhibitive films to be formed but not scale growth which interferes with heat transfer. Other undesirable ions such as Cu, Si, Cl are often removed also from feed water. On the other hand soft waters, which are poor scale formers, contain free carbonic acid which is aggressive. Organic acids from moors and peats also make soft waters aggressive to steel. Such soft waters may have to be hardened by treatment with lime to a pH slightly on the alkaline side of neutral. Some inhibitors function as both anodic and cathodic inhibitors and of these perhaps some of the best known are the polyphosphates. In systems which are at higher temperatures, phosphates and silicates may be used. Temporary protectives such as lanolin and various oils are often used in workshops to protect machines and machined parts and problems can arise when these and other inhibitors are destroyed by bacterial action (Chapter 5).

The Distribution of Corrosion and Situations of Localized Attack

If corrosion becomes concentrated in a localized area of a component the depth of penetration in that area may become very great leading to complete failure of the material. There are many examples of such situations and these include differential aeration, crevice corrosion, pitting, cracking, intergranular corrosion etc. In the present case we shall discuss differential aeration and pitting since these are most often associated with bacterial corrosion and since intergranular corrosion, stress corrosion and corrosion fatigue are adequately covered elsewhere[3,12]

Localized Attack, Differential Aeration and Corrosion under Conditions of Partial Immersion

Various geometrical factors may give rise to regions which are predominantly anodic or cathodic. These may be caused by the presence of dissimilar metals in contact where one is much more active than the other. Differences in metallurgical structure or metallurgical condition due to heat or mechanical treatments or breaks in oxide scales or protective coatings can lead to heterogeneities which in turn may lead to enhanced localized attack. Variations in solution velocity, pH or salt concentration, or in aeration or concentration of cathodic depolarizers may also lead to conditions of enhanced localized attack. Evans[57,58] pointed out that any geometrical factor which resulted in a high oxygen concentration in one area of a metal surface and a low one at another would bring into being cathodic and anodic areas, respectively, on the metal surface. This is known as the Differential Aeration Principle and differential aeration cells are one of the commonest causes of intense localized attack. The principles of differential aeration have been studied by Pourbaix[59] and this work should be consulted if more detailed explanations are required.

Tuberculation

This type of attack is encountered in a variety of situations and may be encountered in relation to bacterial corrosion (Chapter 6). In ordinary corrosion processes this often occurs at the break or discontinuity in an oxide scale (Fig. 2.12). At the site of the break in the film we have the situation of a large cathode (oxide scale) which can easily sustain oxygen reduction and a small anode which leads to an intense form of corrosion at the bare metal. Anodic dissolution takes place and as metal ions concentrate in the solution the solubility product of the solid hydroxide is exceeded locally and hydroxide precipitates out as a hemispherical membrane which surrounds and covers the original discontinuity. This results in effective screening of the anodic area from available oxygen and the metal at the discontinuity remains anodic. The detailed situation in the case of tuberculation associated with a break in

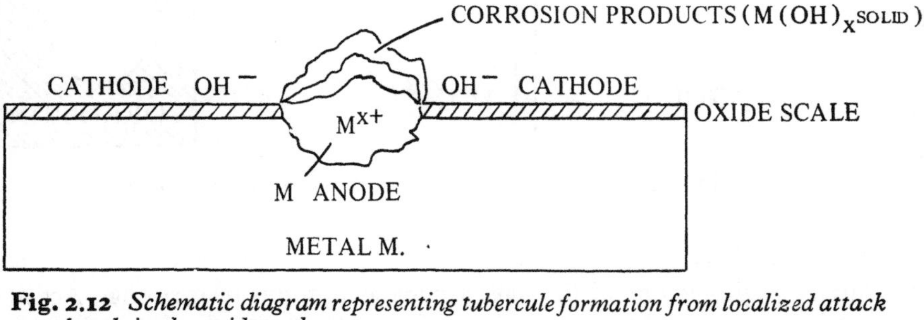

Fig. 2.12 *Schematic diagram representing tubercule formation from localized attack at a break in the oxide scale*

the millscale on a steel pipe is more complex and has been discussed in detail elsewhere[60].

Differential Aeration under Conditions of Partial Immersion

If a metal such as iron or zinc is partially immersed in a neutral salt solution (cf. Fig. 2.13) then corrosion will take the form illustrated. On the upper region of the metal surface oxygen diffusion is relatively easy and this area will become cathodic with respect to the lower part which becomes anodic due to oxygen starvation. In these areas metal passes into solution forming metal chloride whereas in the cathodic areas hydroxyl ions are produced forming NaOH in the case of an NaCl solution. These reaction products then diffuse away from the points on the surface at which they are produced and interact below the water line to form hydroxides in the case of iron ($Fe(OH)_2$ initially) or in the case of zinc $Zn(OH)_2$ and basic chloride $Zn(OH)_2{\cdot}ZnCl_2$. Further reactions may then take place in the case of iron where black or green hydrated magnetite may be formed and if sufficient oxygen is present oxidation to ferric iron and the formation of rust [$FeO(OH)$ or $Fe_2O_3{\cdot}H_2O$] may occur. When the geometry of the system is such as to give rise to a three phase boundary between the metal, electrolyte and atmosphere, and when the oxygen concentration in the atmosphere is low intense water-line corrosion can occur[61,62]. If oxygen is replaced by nitrogen or an inert gas in these situations these forms of localized attack no longer occur.

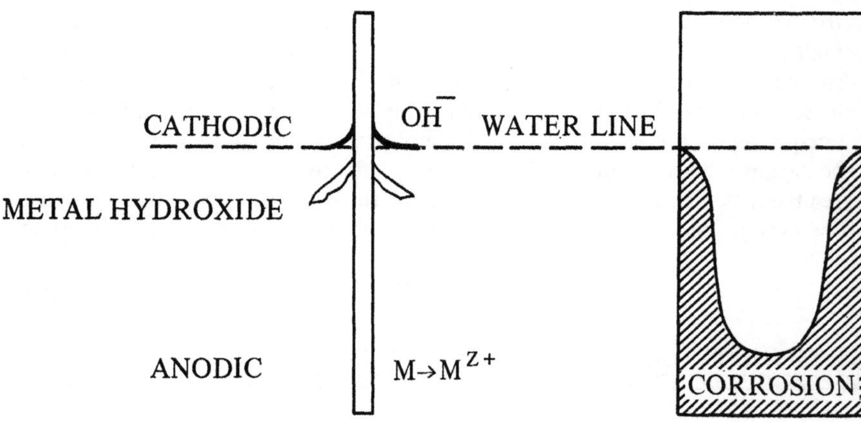

Fig. 2.13 *Corrosion under conditions of partial immersion*

Pitting Corrosion

Surface pitting is one of the most troublesome forms of corrosion. Active metals, *e.g.* Cr and Al, are most susceptible to this form of attack and alloys containing these metals, which are dependent on Cr_2O_3- or Al_2O_3-rich oxide films for their resistance to corrosion, are also susceptible, *e.g.* stainless steels.

The theory and mechanism of pitting corrosion is still the subject of dispute between various researchers[63] although there is considerable support for the catalytic theory[63-65]. Here it is suggested that in chloride solutions the chloride ion migrates to the anodic areas and causes a decrease in pH which may be sufficient to allow metal to dissolve as a soluble salt. The dissolution of metal in turn creates acidity and the process is autocatalytic. It will be remembered that it was indicated earlier that for certain metals, *e.g.* Fe, Ni, Sn, Zn and Al, some initial production of anodic acidity was required for anodic dissolution to take place. This mechanism of pitting of Al has been studied in chloride solutions as has the mechanism of pitting of stainless steels in $FeCl_3$.[65] Generally the agents which are most active in producing pitting are Cl^-, Br^-, I^- and hypochlorite and solutions of oxidizing salts containing $CuCl_2$ and $FeCl_3$.[66] Nitrates may be used as pitting inhibitors for stainless steels in $FeCl_3$ but the inhibitor *must* be present from the start of the exposure otherwise additions later may stimulate pitting[63].

It is believed that on materials of the type referred to above there are no preferential pit sites[67] although these depend for their formation upon irregularities and defects in the protective oxide film. In the majority of cases the pits are hemispherical in shape with an electropolished inner surface. It is considered[68] that the potentials of the anodic and cathodic areas of a metal undergoing pitting differ greatly due to the presence of a film of solution of high electrical resistance at the pit/solution interface. This resistance arises since the anodic products formed in the pit cannot diffuse away rapidly. The electropolishing process in the pit keeps the current density constant and independent of potential. Thus the potential inside the pit is within the active region whilst that outside is within the passive region. The enlargement of the surface area of growing pits occurs with constant increase in diameter per unit time and the corrosion rate inside the pit may be up to 30×10^3 times greater than that on the passive surface. Detailed studies of the pitting corrosion of aluminium in chloride solutions have been made by Edeleanu[69] and a recent discussion of passivation, pitting and polishing should be consulted as a useful general reference[70].

Links between General Corrosion and Microbiological Corrosion

It is hoped that this brief survey of corrosion processes will enable the reader to understand and interpret the five major mechanisms by which microbes, in the broadest sense, bring about corrosion problems. These were indicated in the Introduction to this Chapter and will be considered in detail in the following chapters. Further reading may become necessary and the references cited in the Introduction should cover the necessary ground. More detailed information can be found in the individual papers mentioned in the References. Excellent original papers on corrosion topics appear in a number of well-known journals such as *Corrosion* and the *Journal of the Electrochemical Society* published in the United States, and *Corrosion Science* and the *British Corrosion Journal* published in the United Kingdom. *Werkstoffe und Korrosion* is published in Germany and the Proceedings of the various International Congresses on Metallic Corrosion are also available as reference sources. In addition papers on corrosion will be found in many metallurgical journals published throughout the world. A convenient source of corrosion information are the *Corrosion Abstracts* published by the Swedish Corrosion Institute in Stockholm.

References

[1]BOCKRIS, J. O'M. AND REDDY, A. K. N. *Modern Electrochemistry* (in 2 vols.). New York: Plenum Press, 1970.

[2]DAVIES, C. W. *Electrochemistry*. London: George Newnes Ltd., 1967.

[3]SHREIR, L. L. *Corrosion* (in 2 vols.). London: George Newnes Ltd., 1963.

[4]SPELLER, F. *Corrosion*. London: McGraw-Hill, 1951.

[5]EVANS, U. R. *The Corrosion and Oxidation of Metals*. London: Edward Arnold, 1960.

[6]EVANS, U. R. *The Corrosion and Oxidation of Metals*, 1st Suppl. Vol. London: Edward Arnold, 1968.

[7]UHLIG, H. H. *Corrosion and Corrosion Control*. Chichester: John Wiley & Sons, 1963.

[8]UHLIG, H. H. *The Corrosion Handbook*. London: Chapman & Hall, 1948.

[9]EVANS, U. R. *Introduction to Corrosion*, 2nd edn. London: Edward Arnold, 1963.

[10]SCULLY, J. C. *Fundamentals of Corrosion*. London: Pergamon Press, 1966.

[11]STEWART, D. AND TULLOCH, D. S. *Principles of Corrosion and Protection, Introductory Monographs in Materials Science*. London, Macmillan, 1968.

[12]MENZIES, I. A., POTTER, E. C., EDELEANU, C. AND STANNERS, J. F. *Corrosion and Protection of Metals*, Iliffe Books Ltd., London (1965).

[13]HOAR, T. P. (Ed. Bockris, J. O'M.). '*Modern aspects of Electrochemistry*', **2,** 262 (1959). Butterworths, London.

[14]POURBAIX, M. *Atlas of Electrochemical Equilibrium in Aqueous Solutions*. Oxford: Pergamon Press, 1968.

[15]EVANS, U. R. AND HOAR, T. P. Trans. Far. Soc., **30,** 424 (1934).

[16]HOAR, T. P. Trans. Far. Soc., **33,** 1152 (1937).

[17]EDELEANU, C. AND EVANS, U. R. Trans. Far. Soc., **47,** 1121 (1951).

[18]HOAR, T. P. Trans. Far. Soc., **45,** 683 (1949).

[19]EVANS, U. R. J. Soc. Chem. Ind. (Lond.), **43,** 319t (1924).

[20]HOAR, T. P. AND EVANS, U. R. J. Chem. Soc., 2476 (1932).

[21]HOAR, T. P. Ind. Chim. Belge, **19,** 931 (1954).

[22]BONHOEFFER, K. F. Corrosion, **11**, 304t (1955).
[23]HOAR, T. P. Third Meeting, Intern. Comm. Electrochem. Thermodyn. Kinet., Berne (Manfredi, Milan), 146 (1952).
[24]Reference 3, Chapter 4.6, Volume 1.
[25]EVANS. U. R. J. Franklin Inst., **208**, 52 (1929).
[26]WAGNER, C. AND TRAUD, W. Z. Electrochem., **44**, 391 (1938).
[27]BONHOEFFER, K. F. AND JENA, W. Z. Electrochem., **58**, 151 (1951).
[28]STERN, M., J. Electrochem. Soc., **102**, 605, 663 (1955).
[29]STERN, M. AND GEARY, A. L. J. Electrochem. Soc., **104**, 56 (1957).
[30]STERN, M. Corrosion, **14**, 440t (1958).
[31]STERN, M. AND WEISERT, E. D. Proc. Amer. Soc. Test. Mat., **59**, 1280 (1959).
[32]VONDRÁCEK, R. AND IZÁK, K. J. Trans. Rec. Trav. Chim. Pays-Bas, **44**, 376 (1925).
[33]EDELEANU, C. Metallurgia (Manchr.), **50**, 113 (1954).
[34]EDELEANU, C. AND GIBSON, J. G. Chem. Ind. 301 (1961).
[35]SUDBURY, J. D., RIGGS, O. L. AND SHOCK, D. A. Corrosion, **16**, 47t, 55t (1960).
[36]RIGGS, O. L., HUTCHISON, M. AND CONGER, N. L. Corrosion, **16**, 58t (1960).
[37]LOCKE, C. E., HUTCHISON, M. AND CONGER, N. L. Chem. Eng. Progr., 56 (11), 50 (1960).
[38]MUELLER, W. A. Can. J. of Technol., **34**, 162 (1956).
[39]SUDBURY, J. D., BANKS, W. P. AND LOCKE, C. E. Second Intern. Congr. Metallic Corrosion, New York (1963).
[40]ROBINSON, F. P. A. AND GOLANTE, L. Second Intern. Congr. Metallic Corrosion, New York (1963).
[41]ACELLO, J. AND GREENE, N. D. 19th Annual Conf. N.A.C.E., New York (1963).
[42]COTTON, J. B. Chem. Ind., 1520 (1960).
[43]FISCHER, W. R. Werkstoffe u Korros., **10**, 243 (1959); **17**, 477 (1966).
[44]TOMASHOV, N. D. Corrosion Sci., **4**, 315 (1964).
[45]FOROULIS, Z. A. Corrosion Sci., **5**, 383 (1965).
[46]FOROULIS, Z. A. Ind. engg. Chem. (process design), **4**, 20, 23 (1965).
[47]RIGGS, O. L. Corrosion, **19**, 180t (1963).
[48]Reference 3, Vol. 2, Section 11.
[49]APPLEGATE, L. M. '*Cathodic Protection*', McGraw-Hill, New York (1960).
[50]MEARS, R. B. AND BROWN, R. H. Trans. Electrochem. Soc., **74**, 519 (1938).
[51]HOAR, T. P. Trans. Electrochem. Soc., **76**, 157 (1939).
[52]Report of the National Chemical Laboratory, 1958, H.M.S.O., London (1959).
[53]HOAR, T. P. Chem. Ind., **41**, 118 (1939).
[54]EVANS, U. R. AND BANNISTER, L. C. J. Chem. Soc., 1361 (1930).
[55]Hydrazine Conference Report, Whiffin & Sons Ltd., London (1957).
[56]Reference 3, Section 9.4, Vol. 1.
[57]EVANS, U. R. J. Inst. Metals, **30**, 239 (1923).
[58]EVANS, U. R., BANNISTER, L. C. AND BRITTON, S. C. Proc. Roy. Soc., Ser. A, **131**, 355 (1931).
[59]POURBAIX, M. J. Electrochem. Soc., **101**, 217c (1954).

[60]Reference 3, Section 1.3A, Vol. 1.
[61]HERSCH, P., Nature, **160**, 1407 (1957).
[62]HERSCH, P., Corrosion Technology, **6**, 293 (1959).
[63]GREENE, N. D. AND FONTANA, M. G. Corrosion, **15**, 25t (1959).
[64]HOAR, T. P. Disc. Far. Soc., **1**, 299 (1947).
[65]EDELEANU, C. AND EVANS, U. R. Trans. Far. Soc., **47**, 1121 (1951).
[66]UHLIG, H. H. Trans. Amer. Soc. Metals, **30**, 947 (1940).
[67]WEIL, K. G. AND MENZEL, D. Z. Electrochem., **63**, 669 (1959).
[68]FRANCK, W. F. Proc. First Intern. Congr. Metallic Corrosion, London, 1961, Butterworth, London (1962).
[69]EDELEANU, C. J. Inst. Met., **89**, 90 (1960).
[70]HOAR, T. P., MEARS, D. C. AND ROTHWELL, G. P. Corrosion Sci., **5**, 279 (1965).

CHAPTER THREE: J. D. A. MILLER & A. K. TILLER

Microbial Corrosion of Buried and Immersed Metal

Background to the Problem

Introduction

That unprotected iron or steel objects in the ground or immersed in water tend to corrode is a truism; the electrochemical mechanisms involved are well understood, and are described in Chapter 2. Nevertheless it is still perhaps not generally realized that a large proportion of this corrosion is brought about, or assisted, by the action of microbes.

For the proper understanding of microbial processes leading to corrosion in these situations, some basic knowledge of the range of physical and chemical conditions in natural environments is necessary. Soil, in particular, is a complex and extremely variable environment; thus it is often difficult to trace the mechanism of underground corrosion, or to separate the purely electrochemical processes from those caused by microbes. This is especially true in situations where cyclical changes in conditions enable first one process and then another to predominate and where, therefore, different types of corrosion product are superimposed on one another or existing ones are altered and become unrecognizable. It has been estimated, though, that in the UK at least 50% of failures of underground pipes are due to microbial action[1], and that as long ago as 1955 the maintenance and replacement of underground pipes because of corrosion cost about £20 million.[2]

The Soil

"Soil", for the purposes of this article, can be taken to mean all types of substratum from sand to clay, from highly acidic and almost purely organic peat with great water retention to the topsoil of downland, consisting sometimes of little more than lumps of chalk and hence retaining scarcely any moisture.

Pure peat apart, soils owe their basic structure to inorganic particles that first arose by the weathering of the rock of the earth's mantle. These particles have been sorted out by the action of wind and water, and according to their situation and their chemical nature have been colonized to a greater or lesser extent by pioneer plants such as lichens. With the death and decay of these early colonizers, and transport and deposition of other plant and animal remains, the primitive soils have acquired a variable amount of organic material (least so in the constantly washed sand on the shores of seas and rivers) which has provided the nutriment for an often rich and varied microbial population; and some type of covering of "higher" vegetation has developed, characteristic of the water content, pH, fertility etc.

There are thus various possible ways of classifying soils according to mineral particle size, content of organic material, type of vegetation they support, and so on; a convenient classification for our purposes is on the basis of particle size, since this determines some properties of soil which bear upon corrosion.

The mineral particles in soil range from gravel (1–2 mm or more in diameter), through coarse and fine sand (the finest usually being called silt) to clay, the particles of which are generally less than 5 μ in diameter and approach the colloidal range. Soil types derived from these particles range from pure gravel or sand, well-drained and with no organic content, through sandy loam, loams (light and heavy) with high organic content, to clay loam and at the end of the scale pure clay, which if the particles are small enough will be impermeable to water and relatively free from organic matter.

Although the treatment of electrochemical corrosion unassisted by microbial action is outside the scope of this chapter, it should be mentioned in passing – indeed it is clear from the preceding chapter – that this type of corrosion is prevalent in acid soils such as peat, and also in more neutral situations such as coastal and estuarine soils which have a high content of ions. Moreover, all except markedly alkaline soils could be expected to be aggressive to some degree towards buried metal objects provided that both oxygen and moisture are present.

The degree of oxygenation and the water-holding capacity of a soil are interdependent, and are determined by the particle size, larger particles giving a soil with low water-holding capacity and large air capacity. It follows that clays and clay loams, which abound over much of the UK owing to the Pleistocene glaciation, have a high water-holding capacity and readily become waterlogged. If there is a significant organic content in such clayey soils, supporting the growth of a microbial population, oxygen (solubility in water about 10 parts per million) will rapidly be removed from the water in the soil since the rate of consumption by the microflora and microfauna will almost always exceed the rate of dissolution of oxygen at the surface and transport downwards by diffusion. Oxygen will only return to this soil by its drying out or by mass movement of oxygenated water into it. The process of deoxygenation is also described in Chapter 1 under *Biochemical Oxygen Demand.*

Clay soils are frequently near neutral in reaction (pH about 6–8), favouring the growth of micro-organisms. The common presence of water in these soils might also be thought to favour microbial growth; though in fact the well-drained soils are usually quite damp enough, at least in temperate countries, to support an active microbial population.

Water

A similar situation to the waterlogged clay soils can exist at the bottom of rivers and lakes. In the upper reaches of rivers, the bed is scoured free of loose material by the rapid current of water which is itself well aerated by its churning motion; but in the lower reaches the current is slower and deposition of this scoured material begins to occur. This deposited material may remain scarcely disturbed, especially at and near estuaries. The same applies, of course, to the bottom mud of lakes and ponds. Oxygen-free conditions can then very easily arise in these bottom muds by the action of aerobic and micro-aerophilic microbes; and frequently the situation is exacerbated by the water itself having a high BOD (see Chapter 1) because of pollution, and supporting heavy microbial growth.

Types of Microbial Corrosion

In this chapter we are going to consider microbial corrosion of metals in a variety of natural conditions. For micro-organisms to produce their characteristic chemical effects they usually must be growing, not merely lying dormant (though secondary, physical effects can be produced merely by the *presence* of a thick clump of micro-organisms, whether growing or not). This would suggest that, in general, damp or wet environments of near-neutral pH, well aerated and with a supply of organic nutrients, would most favour the growth of any microbes involved in corrosion. Though this is partly true, the small group of bacteria which cause by far the worst havoc in this respect – the sulphate-reducing bacteria – have different requirements from the above in one important respect: they are anaerobic, and indeed are quite unable to grow in the presence of any but the most minute traces of oxygen. Thus, ironically, the soil conditions which an electrochemist might believe were the ideal ones in which, for example, to lay a pipeline – a neutral clay soil, water-logged and hence oxygen-free for much of the year, so that polarization of the cathodic areas of the buried metal should prevent anodic dissolution – are in fact the ones that most favour the growth of the anaerobic sulphate-reducing bacteria.

It should be pointed out here that the *immediate* cause of corrosion is electrochemical in all the cases we are about to consider. The sulphate-reducing bacteria appear to allow a stifled electrochemical reaction to be resumed, while initiating another; a different group of microbes produces acids, mineral and organic, with well-known corrosive effects; a third forms thick clumps which allow differential aeration and concentration cells to arise; while a fourth, already described very briefly in the legend of Fig. 1.11 in Chapter 1, breaks down protective coatings and thus allows electrochemical corrosion (pure, or microbially-induced) to occur.

Corrosion by the Sulphate-Reducing Bacteria

Historical

It was suspected even before the turn of the century that microbes might have a role in the corrosion of metals immersed in water, and in 1910 Gaines[3] suggested that iron bacteria and sulphur bacteria were responsible in part for the corrosion of buried ferrous metals. By 1924 it was considered[4] that hydrogen sulphide production by sulphate-reducing bacteria may play an important part in this type of corrosion. Von Wolzogen Kühr noticed in Holland that corrosion associated with sulphide occurred in wet, anaerobic soil in which polarization of cathodic areas would be expected to inhibit corrosion. Realizing that there may therefore be other depolarizing agents, that is, hydrogen acceptors, present in the soil besides atmospheric oxygen, he and Van der Vlugt made the first attempt[5] to provide a detailed explanation of underground corrosion by sulphate-reducing bacteria in electrochemical terms. It was already known[6] that sulphate-reducing bacteria were able to take up molecular hydrogen while reducing sulphate according to the equation:

$$SO_4^{2-} + 4H_2 \rightarrow S^{2-} + 4H_2O \qquad (1)$$

They proposed that these bacteria could similarly remove the *atomic* hydrogen from polarized cathodes, oxidizing it to protons and electrons, and utilizing the reducing power so obtained for the reduction of sulphate. Full details of their theory are given further on.

It will now be helpful to look at the properties of the sulphate-reducing bacteria in a little detail.

Characteristics of the Sulphate-Reducing Bacteria

The sulphate-reducing bacteria (or more correctly the anaerobic dissimilatory sulphate-reducing bacteria, since numerous microbes reduce small amounts of sulphate for the synthesis of sulphur-containing organic substances) are abundant in nature, and probably ubiquitous. They have been found in Europe, Asia, Africa, North and South America, Australasia and Antarctica; and they have been encountered at a depth of 71m in clay. In the UK it is a rare occurrence to take a soil sample in which they cannot be demonstrated. In this connexion it is difficult to tell whether a soil is oxygen-free *in situ*, because disturbance by taking a sample or even by pushing an electrode into it to measure redox potential inevitably leads to some oxygenation; but nevertheless open-textured dry soils which must undoubtedly be oxygenated will almost always contain demonstrable sulphate-reducing bacteria. The organisms cannot be active in these situations, but they are nevertheless viable, so that the onset of favourable conditions in the soil will soon result in their growth.

Water, both fresh and salt, often contains sulphate reducers, especially if the BOD is high. Whereas water-living organisms usually tend to be physiologically adapted either to fresh-water or salt-water habitats but not both, most strains of the sulphate-reducing bacteria hitherto isolated are readily adaptable between NaCl-free medium and that containing the sea-water concentration, with the exception of a few strains of marine origin (members of the species *Desulfovibrio salexigens*) which appear to have an absolute requirement for the chloride ion. There is some confusion in the terminology used by microbiologists here: microbes isolated from a marine environment (whether or not they can be adapted to salt-free conditions) have often been described as "halophilic", while some microbiologists would reserve this term for organisms found growing in more strongly saline habitats such as salt lakes in desert regions and would describe isolates from the sea as "marine" or "salt-water" forms. In this review, in common with many of the authors we have cited, we use the term *halophile* for any laboratory strain that originated from salt water.

Certain features of their metabolism suggest that the sulphate-reducing bacteria are very ancient, or "primitive" inasmuch as they may be an ancestral form that has given rise to many of the species of living things existing today: they certainly must have found the early conditions on this planet, such as the reducing atmosphere, highly conducive to active growth.

It is doubtful whether until fairly recently many bacteriologists had pure cultures of sulphate-reducing bacteria; this is borne out by certain activities attributed to these organisms in the earlier literature which cannot now be demonstrated. There are, though, a number of distinct species of sulphate reducers (eight are now recognized, some of these having aberrant strains or distinct varieties), so that some of the conflicting results of earlier days may be explained by the unawareness of workers in different laboratories that they were handling different types. Even today, much work remains to be done in comparing the nutrition, physiology and biochemistry of these newly recognized species. Moreover, even with pure cultures and refined techniques at our disposal, we are still perhaps not aware of, or do not fully understand, all the chemical processes involved in corrosion by these bacteria, and are certainly not yet able to assess the relative contributions of the various known processes to the overall reaction.

Not the least of the desulfovibriologist's problems is merely that of growing the bacteria: even experienced workers sometimes encounter complete failures of growth. In addition, laboratory studies are rendered difficult by certain peculiarities of these organisms. Examples are their production of sulphide ions, which react with the ferrous ions that must be incorporated in the growth medium, forming a precipitate or fine suspension of ferrous sulphide; and their excretion, especially in later stages of growth and more especially by halophiles, of a viscous mucin in which a considerable number of the bacteria become embedded. Population densities and growth rates are thus impossible to determine accurately, and uniform or standardized suspensions of the bacteria are difficult to prepare; while in electrochemical studies the electrodes may be poisoned by the sulphide so that readings are unreliable.

* * *

A table showing the modern classification of the sulphate-reducing bacteria, with synonyms, is given below (Table 3.1); it will be seen that all

Table 3.1 *List of the sulphate-reducing bacteria, with some of their more important characteristics*

Species	Synonyms (old names)	Characteristics
Genus *Desulfovibrio*	*Vibrio, Sporovibrio, Spirillum, Microspira*	Single flagellum, or a tuft. Do not form spores. Hydrogenase usually present·
Dv. desulfuricans	*Desulfovibrio desulfuricans*	Curved rods (vibrios); sometimes spirilloid, occasionally straight. Typical size 3–5 μ × 0·5–1 μ. Single flagellum. Growth on pyruvate or choline in absence of sulphate.
Dv. vulgaris	*Desulfovibrio desulfuricans*	Curved rods (vibrios); sometimes spirilloid, occasionally straight. Typical size 3–5 μ × 0·5–1 μ. Single flagellum. Growth on pyruvate or choline in absence of sulphate.
Dv. salexigens	*Desulfovibrio desulfuricans*	Curved rods (vibrios); sometimes spirilloid, occasionally straight. Typical size 3–5 μ × 0·5–1 μ. Single flagellum. Obligate salt-water species (requires Cl^-).
Dv. gigas		Large curved rods or spirilla, 5–10 μ × 1·2–1·5 μ. Tuft of flagella.
Dv. africanus		Long, slender, sigmoid rods, 5–10 μ × 0·5 μ. Tuft of flagella.
Genus *Desulfotomaculum*		Peritrichous flagella (*i.e.*, distributed all over the organism). Spores formed.
Dt. nigrificans	*Vibrio thermodesulfuricans, Clostridium nigrificans*	The only thermophilic species (optimum temperature 55 °C). Rods, 3–6 μ × 0·3–0·5 μ. Hydrogenase activity variable; not coupled to sulphate reduction. Growth on pyruvate in sulphate-free media.
Dt. orientis	*Desulfovibrio orientis*	Fat curved rods, 5 μ × 1·5 μ. Hydrogenase apparently absent.
Dt. ruminis		Rods, 3–6 μ × 0·5 μ. Growth on pyruvate in sulphate-free media. Found in rumen of sheep.

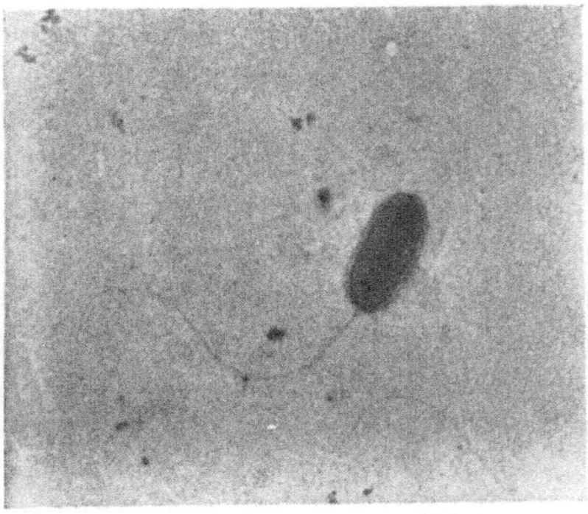

Fig. 3.1 *Electron micrograph of* Desulfovibrio desulfuricans, *showing single flagellum* (× 5000)

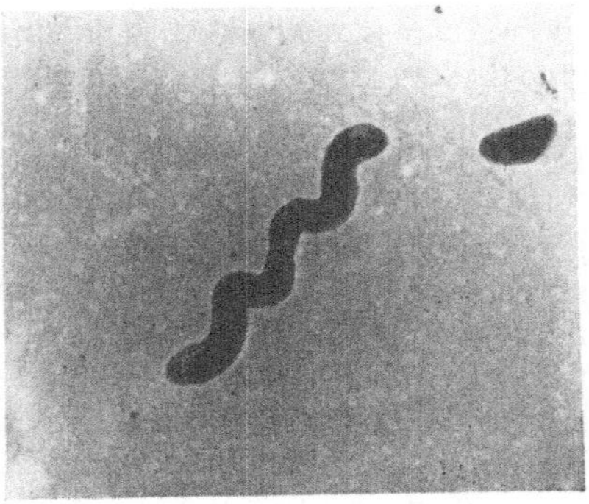

Fig. 3.2 *Electron micrograph of a spirillar form and a swollen form of* Desulfovibrio desulfuricans (× 3000)

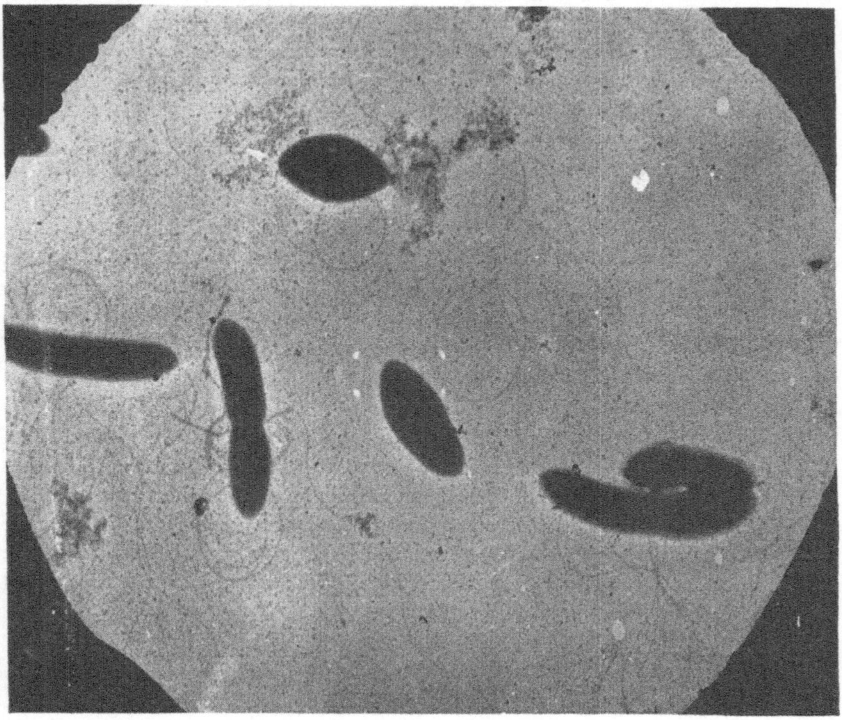

Fig. 3.3 *Electron micrograph of* Desulfotomaculum nigrificans, *showing numerous flagella and swollen forms* (× 10 000)

species except one are mesophilic, that is, their optimum temperature for growth lies within the normal limits of temperate and tropical climates (say 10–40°C). Some electron micrographs of sulphate reducers are shown in Figs. 3.1–3.3.

Table 3.1 also shows that there are marked dissimilarities between the genera *Desulfovibrio* and *Desulfotomaculum*. The difference in the number and arrangement of the flagella in the two genera would alone be enough to make many bacterial taxonomists separate them widely, and there is no doubt that they are biochemically very different though, as already stated, many more studies need to be made. We shall be concerned principally with members of the genus *Desulfovibrio* in this article.

Desulfovibrio spp. in pure culture in the laboratory can use sulphate, sulphite and a few other inorganic sulphur-containing anions as terminal electron acceptors for energy-yielding reaction;[7] that is, they carry out a form of anaerobic respiration. They are chemo-organotrophic heterotrophs, though they appear to grow well on relatively few organic carbon-and-energy sources. Salts of certain carboxylic acids (lactic, pyruvic, malic, fumaric) support the most vigorous growth, while markedly poorer growth is obtained with a few sugars, alcohols etc. One strain utilizes oxamate; certain others will grow on pyruvate, fumarate or choline in sulphate-free medium by a fermentative type of reaction (see Chapter 1). All the strains that have been examined appear to grow normally under nitrogen, so that hydrogen is not essential for growth reactions. Some strains can obtain energy by the oxidation of certain simple organic substances such as oxamate, mentioned above, which they cannot use as carbon sources. Earlier workers claimed that *Desulfovibrio* is chemolithotrophic, using the oxidation of hydrogen to provide energy for the autotrophic assimilation of CO_2; but recent reports, and unpublished work in the Manchester laboratories, cast strong doubts on the existence of autotrophy and chemolithotrophy respectively. The nutrition of the sulphate reducers, and many other aspects, have been dealt with by Postgate in a very valuable review.[8]

In most of the recent laboratory studies, sulphate reducers have been grown in media containing sodium lactate as a carbon source. Yeast extract is usually added, since this results in more rapid and abundant growth, though the few strains that have yet been critically examined can grow in the absence of yeast extract in a mineral salts + sodium lactate medium. Acetate and CO_2 are the major end-products of the growth reaction, the overall equation for which can be written:

$$2CH_3.CHOH.COONa + SO_4^{2-} \rightarrow 2CH_3.COONa + 2CO_2 + 2H_2O + S^{2-} \quad (2)$$

This can be expressed in a convenient pictorial form often used by biochemists, to illustrate the linked cell reactions:

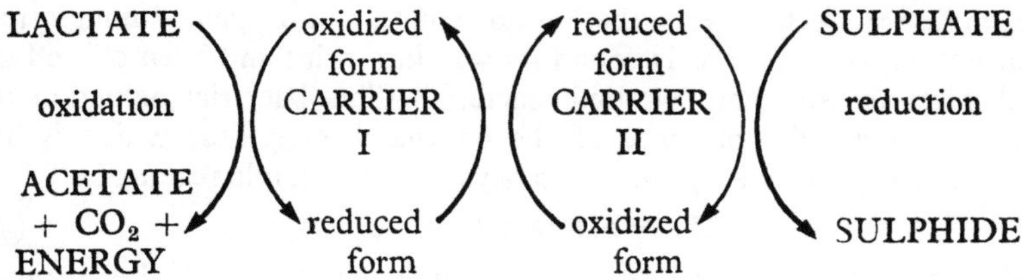

(3)

There are thus several intermediate stages in this process, involving the cyclical reduction and re-oxidation of intermediate electron carriers (two are shown participating in the reaction), but these need not concern us here. The important point is that the E'_0 for the initial oxidation (the lactate system) must, just as in any non-biological electrochemical coupled reactions, be more negative than that for the final (sulphide) system.

We can therefore simplify scheme (3) to the following:

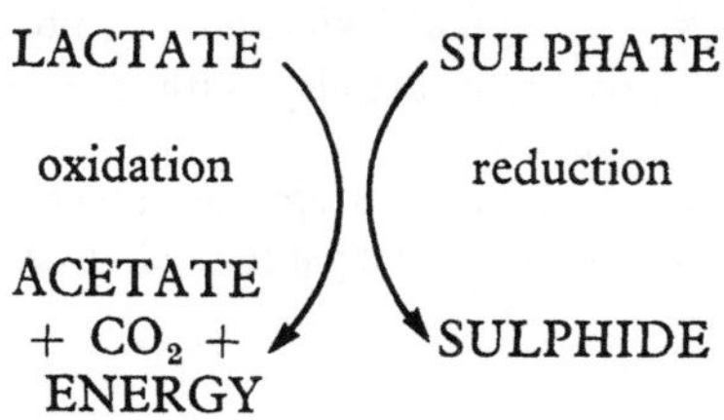

(4)

The analogy with aerobic respiration – the standard type of oxidation of a sugar by the majority of living things, using atmospheric oxygen as the terminal electron acceptor – can now easily be shown:

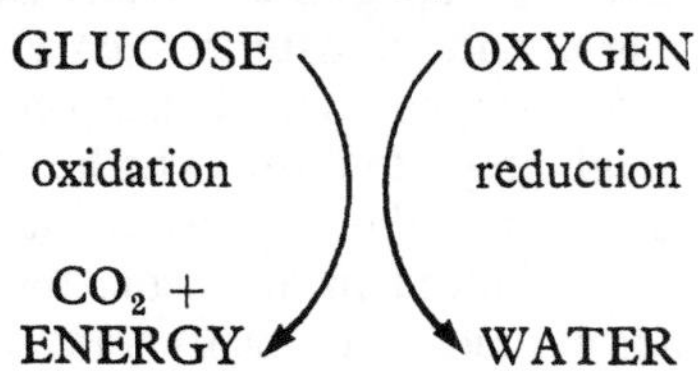

(5)

Equation (1) can of course similarly be written:

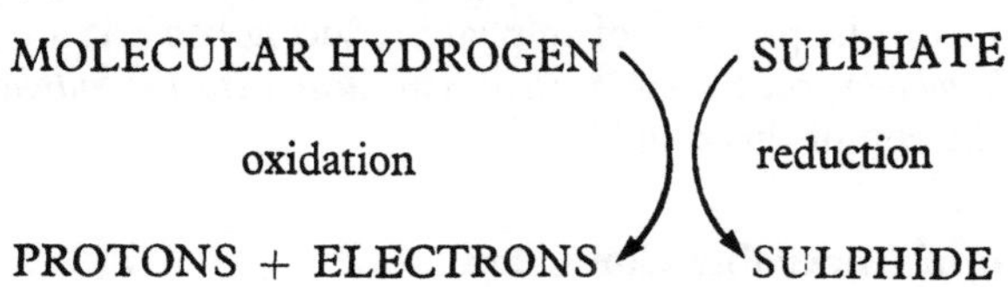

(6)

(the oxygen atoms from the sulphate radicals combining of course with the hydrogen to form water: hence there is no evolution of oxygen, although this misconception has occasionally found its way into print and been offered as a mechanism of corrosion by these bacteria). This bacterial oxidation of hydrogen requires the presence of the enzyme *hydrogenase*, which is an unusual enzyme in that it appears to catalyse two reversible reactions:

$$H_2 \rightleftharpoons 2H \rightleftharpoons 2H^+ + 2e \qquad (7)$$

Hydrogenase activity towards *molecular* hydrogen can easily be demonstrated in non-growing suspensions of sulphate-reducing bacteria in phosphate or other buffer at about neutral pH and under oxygen-free hydrogen, either by adding sodium sulphate solution and measuring the hydrogen uptake manometrically or adding an oxidation-reduction dye such as methylene blue or benzyl viologen and observing its colour change. The hydrogen uptake in μl per standard amount of bacteria (1 mg dry weight) per hour determined manometrically is called the hydrogen absorption coefficient, $-Q_{H_2}$, and it should always be made clear which electron acceptor was used, *e.g.*, $-Q_{H_2}^{SO_4{}^{2-}}$ or $-Q_{H_2}^{BV}$, (= benzyl viologen: sulphate reducers frequently show much greater hydrogenase activity with the redox dyes as electron acceptors than with sulphate).

Sulphate reduction to sulphide is a stepwise process catalysed by a number of enzymes that tend to be collectively referred to as the *sulphate reductase system.*

* * *

The extreme sensitivity of sulphate-reducing bacteria to oxygen means that growth tests designed to establish whether these bacteria are present in a sample of soil, water, corrosion product etc., or to obtain a most probable number, can be unreliable in inexperienced hands. In rich media, or in the presence of a large amount of natural material such as a pellet of soil, growth will often occur in a screw-capped vessel completely filled with freshly boiled-out medium, though it is doubtful whether this growth originates from more than a few per cent of the sulphate reducers initially present. In media used for counts, and indeed in all media for laboratory work with these bacteria, it is standard practice to incorporate reducing agents such as sodium sulphide, or neutralized cysteine hydrochloride, thioglycollic acid or ascorbic acid, to ensure that all or most of the viable organisms in the inoculum will grow.

Actual tests for the presence of sulphate-reducing bacteria are given in the appendix *Microbiological Examination of Soil, Water etc. for Sulphate-Reducing Bacteria* at the end of this chapter.

Mechanism of Anaerobic Corrosion

Support for the Cathodic Depolarization Theory

As we have already seen, the sulphate-reducing bacteria had long been known to reduce sulphate to sulphide using molecular hydrogen (equation (1)). Von Wolzogen Kühr and Van der Vlugt[5], studying the corrosion of buried pipes in the polder districts of Holland – low-lying areas of clayey soil, rich in organic matter and often waterlogged – noticed that iron sulphide occurred, not only as an adherent corrosion product on the pipes themselves but also in the soil in the vicinity of these corroding pipes. From its presence they deduced, and duly confirmed, the presence of sulphate-reducing bacteria; and they suggested that the bacteria acted as a cathodic depolarizing agent, stimulating the corrosion process by the removal of the (atomic) cathodic hydrogen which was then utilized for the reduction of sulphate to sulphide. The overall mechanism they postulated is as below.

anodic reaction	$4Fe \rightarrow 4Fe^{2+} + 8e$	(8)
electrolytic dissociation of water	$8H_2O \rightarrow 8H^+ + 8OH^-$	(9)
cathodic reaction	$8H^+ + 8e \rightarrow 8H$	(10)
cathodic depolarization by bacteria (compare equation (1))	$SO_4^{2-} + 8H \rightarrow S^{2-} + 4H_2O$	(11)
corrosion product	$Fe^{2+} + S^{2-} \rightarrow FeS$	(12)
corrosion product	$3Fe^{2+} + 6OH^- \rightarrow 3Fe(OH)_2$	(13)
overall reaction	$4Fe + SO_4^{2-} + 4H_2O \rightarrow 3Fe(OH)_2 + FeS + 2OH^-$	(14)

It follows from these equations that the ratio of iron corroded to FeS produced (mole:mole) should be 4:1, but in fact values from 0·9 to 48 have been recorded.

Other evidence has been presented, both for and against the "Dutch theory". Wanklyn and Spruit[9], who observed high ratios of iron corroded/FeS produced, reported an *anodic* depolarization. They concluded that corrosion was due to anodic stimulation by sulphide produced by bacteria utilizing organic electron donors, and considered that the consequent fall in the anode potential would increase the e.m.f. for hydrogen evolution at the cathode. Other work has broadly supported these findings[10,11]. On the other hand, early attempts to confirm the Dutch theory experimentally by demonstrating cathodic depolarization by sulphate-reducing bacteria were inconclusive.

Eventually, in 1959, Horváth and Solti[12] studied the polarization curves for mild steel electrodes in cultures of sulphate reducers and showed that the strain they used did bring about cathodic depolarization. Although this provided strong evidence in favour of the Dutch theory it lost some of its force because of inadequate characterization of the organism.

An independent study by Booth and Tiller[13] using pure cultures of sulphate-reducing bacteria also showed quite conclusively that cathodic depolarization by the Hildenborough strain of *Desulfovibrio vulgaris* could be effected, but only when the culture was in active growth. This organism is hydrogenase-positive; a hydrogenase-negative organism (*Desulfotomaculum orientis*) did not cause cathodic depolarization. However, both organisms brought about an initial anodic stimulation, followed by an inhibitory effect due to the presence of a partially protective film of ferrous sulphide that gradually formed over the surface. The polarization cell they used is shown in Fig. 3.4.

Supporting these results were those of Booth and Wormwell[14] who studied the rate of corrosion of mild steel in batch cultures of various pure strains of non-halophilic sulphate reducers of different $-Q_{H_2}^{SO_4^{2-}}$. They found a linear relationship between corrosion rate and hydrogenase activity towards sulphate, and concluded that the ability to bring about cathodic depolarization was the most important factor in the stimulation of corrosion by sulphate-reducing bacteria. (As will be seen below, these results were not confirmed when halophiles were examined, nor in semicontinuous and continuous culture.)

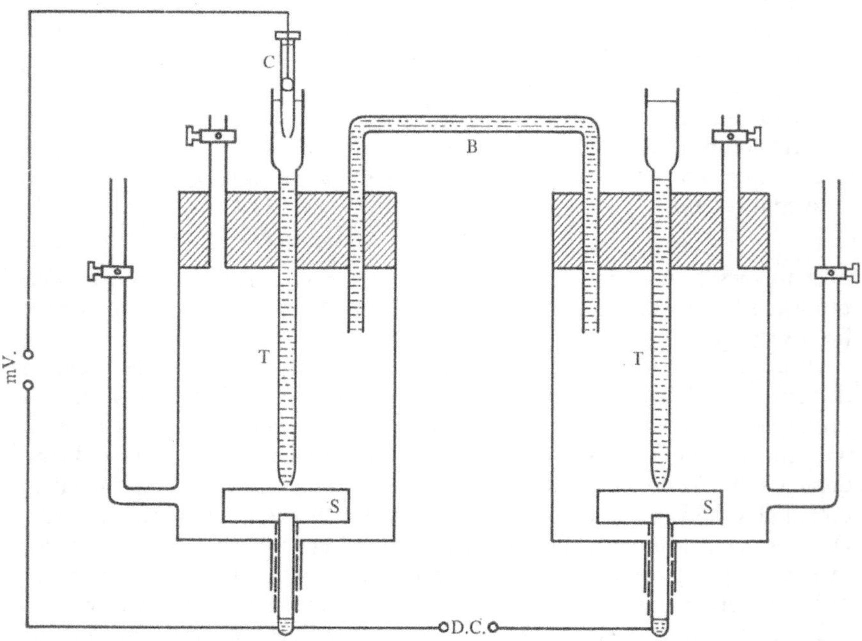

Fig. 3.4 *Divided polarization cell.* The electrodes (mild steel specimens, S) are mounted horizontally in glass vessels which have provision for deaeration by a stream of nitrogen. Electrical contact between the solutions in the vessels is *via* a KCl solution/agar bridge (B); the mild steel electrodes are connected to the galvanostatic circuit. The potential of an electrode is determined with respect to a saturated calomel electrode (C), using a KCl/agar tubulus (T) as a Luggin probe. (By courtesy H.M.S.O. Crown Copyright)

Extension of the polarization studies to include thermophilic strains enabled a fresh assessment of the mechanism of depolarization to be made[15]. Two strains of *Desulfotomaculum nigrificans*, both hydrogenase-negative towards sulphate, behaved differently in polarization cells: one, which was hydrogenase-*positive* towards benzyl viologen, caused cathodic depolarization, while the other, hydrogenase-negative towards the redox dye, did not depolarize. Both strains caused ennoblement of the anode by formation of a sulphide film.

The strain possessing the enzyme hydrogenase was apparently unable to couple its action to that of the sulphate reductase system, suggesting the absence of an essential electron carrier (see linked-reaction scheme (3)). Such a carrier could be the red pigment cytochrome c_3, present in *Desulfovibrio* but absent from *Desulfotomaculum*. The reduction of a redox dye by hydrogenase-catalysed oxidation of hydrogen needs no intermediate electron carrier. Hydrogenase activity was assumed, though, to be occurring in the polarization cell in which there was cathodic depolarization but in which of course no redox dye was present. It was therefore considered that either some organic intermediate metabolite was serving as electron acceptor for hydrogen oxidation, or that hydrogenase was performing only the first reaction in equation (7) and in the direction $2H \rightarrow H_2$.

Four halophilic types were next studied, in saline medium[16]. One, El Agheila Z, was a strain of the species *Desulfovibrio* (*Dv.*) *desulfuricans* (in the new sense – see Table 3.1) which was hydrogenase-negative towards both sulphate and benzyl viologen. A hydrogenase-positive variant of this strain towards both electron acceptors had been produced by growing it under hydrogen with repeated subculturing for a year; this was also studied for depolarizing effects. In addition, two obligate halophiles of the species *Dv. salexigens* were examined.

The results of these experiments, and those on *Dv. vulgaris*, *Desulfotomaculum* (*Dt.*) *orientis* and the thermophiles, are all presented for comparison in Table 3.2.

Anodic and cathodic depolarization curves for mild steel in sterile saline medium are shown in Fig. 3.5, and in presence of the actively growing strains El Agheila Z and Z_1, and the *Dv. salexigens* strains, in Figs. 3.6–3.9.

Several interesting facts emerged. Firstly, the presence and absence of hydrogenase respectively in the two *Dv. desulfuricans* strains correlated with their power to depolarize the cathode, confirming the previous work. Secondly, though in all cases a black film of ferrous sulphide was laid down on the anodes during the course of the experiment, ennoblement of the anode potential did not occur in the case of *Dv. salexigens* strain California 43:63 (Fig. 3.9). In other words, the anodic sulphide film produced during the growth of this strain appeared to be relatively unprotective, though no chemical difference between this film and any of those produced by other strains could be detected. Thirdly, the extreme vigour with which strain California 43:63 caused cathodic depolarization correlated better, as in

Table 3.2 *Hydrogenase and depolarization activity of some sulphate-reducing bacteria*

Organism	N.C.I.B.* Number	Hydrogenase Activity with Sulphate	Hydrogenase Activity with Benzyl Viologen	Cathodic effect	Anodic effect	Reference
Fresh-water mesophiles						
Desulfovibrio vulgaris strain Hildenborough	8303	++	+†	Depolarization	Initial stimulation of anodic reaction, followed by a move to a more noble potential	13
Desulfotomaculum orientis strain Singapore 1	8382	−	−†	No depolarization		
Thermophiles						
Desulfotomaculum nigrificans strain Teddington Garden	8351	−	+	Depolarization	Formation of partially protective sulphide film	15
Desulfotomaculum nigrificans strain Staines Garden	8353	−	−	No depolarization		
Halophilic mesophiles						
Desulfovibrio desulfuricans strain El Agheila Z	8380	−	−	No depolarization	Initial stimulation of the anode reaction followed by a move to more noble potentials	16
Desulfovibrio desulfuricans strain El Agheila Z_1	8380/1	+	++	Depolarization		
Desulfovibrio salexigens strain El Agheila C	8308	+	+	Depolarization		
Desulfovibrio salexigens strain California 43:63	8364	+	++	Vigorous depolarization	No ennoblement: sulphide film evidently non-protective	

Divided polarization-cell technique (separate anode and cathode compartments).
Experiments lasted for varying periods of up to 11 days after inoculation.
Medium for halophiles contained 2·5% sodium chloride.
Thermophiles incubated at 55°C; mesophiles at 30°C.
*National Collection of Industrial Bacteria: see *Appendix*.
†Professor J. R. Postgate: unpublished results.

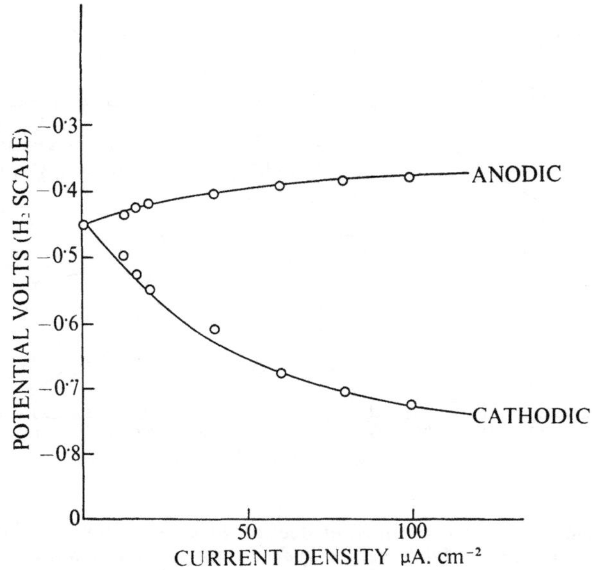

Fig. 3.5 *Polarization curves for mild steel at 30°C in sterile saline medium (Crown Copyright reserved)*

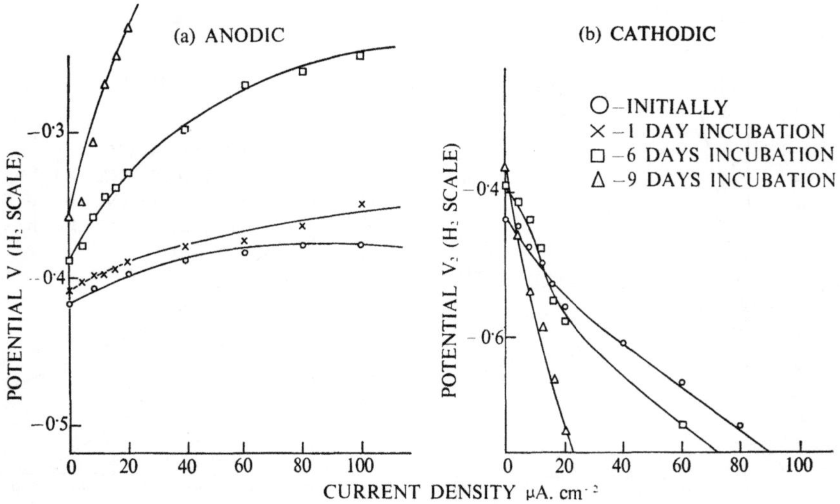

Fig. 3.6 *Polarization curves for mild steel at 30°C in cultures of* Desulfovibrio desulfuricans *(strain El Agheila Z, a hydrogenase-negative strain), indicating both the formation of a protective sulphide film and a lack of cathodic depolarization by the organisms (Crown Copyright reserved)*

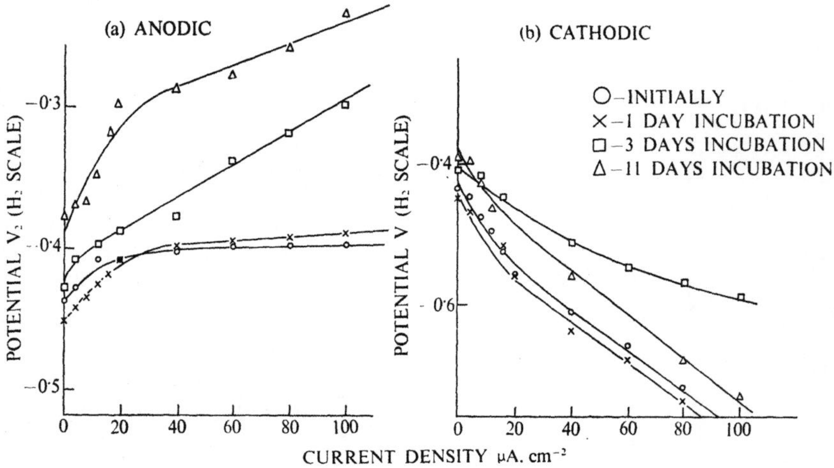

Fig 3.7 *Polarization curves for mild steel at 30°C in cultures of* Desulfovibrio desulfuricans *(strain El Agheila Z_1, a hydrogenase-positive strain), showing again the formation of a protective sulphide film, and cathodic depolarization during active growth (Crown Copyright reserved)*

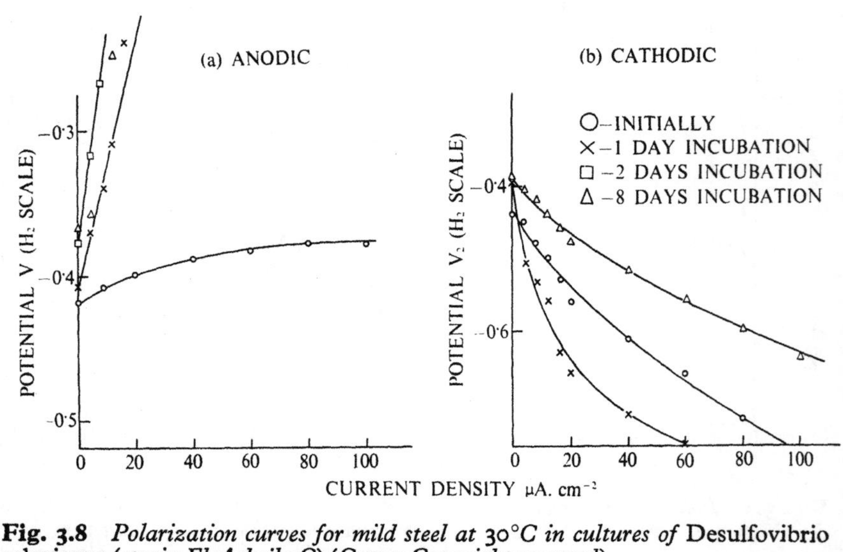

Fig. 3.8 *Polarization curves for mild steel at 30°C in cultures of* Desulfovibrio salexigens *(strain El Agheila C) (Crown Copyright reserved)*

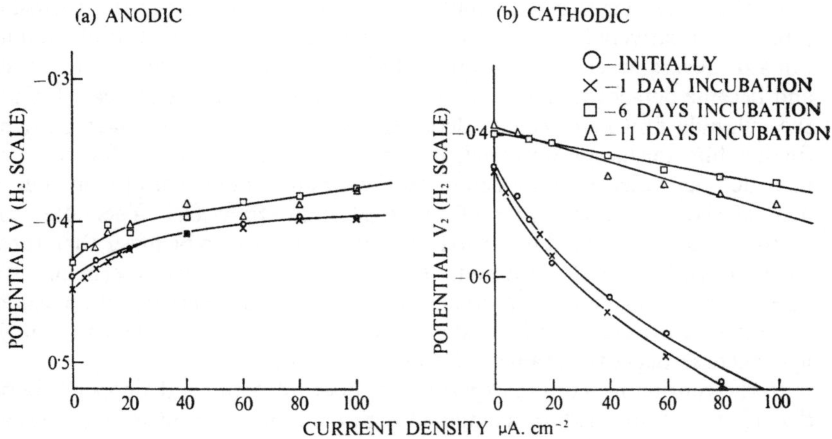

Fig. 3.9 *Polarization curves for mild steel at 30°C in cultures of* Desulfovibrio salexigens (*strain California* 43:63) (*Crown Copyright reserved*)

previously examined strains, with its $-Q_{H_2}^{BV}$ than its $-Q_{H_2}^{SO_4^{2-}}$; in fact, no hydrogenase activity by this strain towards sulphate has ever been demonstrated in the Manchester laboratories.

Booth and Tiller concluded that the ability of a sulphate-reducing bacterium to utilize hydrogen, for whatever purpose, is the criterion for anaerobic microbial corrosion of ferrous metals, and that the earlier equation of corrosion with sulphate reduction is unwarranted.

Recent Developments

Although the work just described undoubtedly confirmed that the sulphate-reducing bacteria play an important role in the corrosion of ferrous metal, the *rate* of corrosion in all these experiments was low compared to those often reported in metal objects exhumed from natural situations; this was attributed to the small amount of chemical activity of the bacteria in batch culture experiments in which there is only a limited supply of nutrients. This objection was met to some extent by determining the rates of corrosion of mild steel in semicontinuous[17] and continuous[18] culture. (Semicontinuous culture is a technique in which a portion of a culture is run off aseptically at intervals and replaced by an equal volume of sterile growth medium. If this operation is designed to give a long mean residence time, for example by removing one-fifth of the culture daily and topping up[17], the organisms grow out of the exponential phase of growth into the stationary phase – even at the unusually slow rate at which the sulphate-reducing bacteria grow – and are subsequently "rejuvenated" by each addition of medium. In true continuous culture, on the other hand, the organisms rapidly reach a steady state).

In the early stages of semicontinuous culture experiments the corrosion rates were relatively low owing to the formation of a protective sulphide film, but after several weeks of exposure to the bacterial action these films fractured and became detached, and the corrosion rate then markedly increased. Once the sulphide film had become detached, either wholly or in fragments, no further film formation occurred. The time interval before the film became detached depended to some extent on the strain of organism present, but in general the detachment process began after 20–30 weeks. Thereafter the corrosion rated increased up to sixfold, but still tended to be lower than those encountered in natural conditions. In these semicontinuous experiments the direct correlation between corrosion and hydrogenase activity disappeared, although hydrogenase-positive strains were more corrosive than the single hydrogenase-negative organism (*Dt. orientis*) tested.

In the continuous culture experiments, Booth, Cooper and Cooper[18] found that if the culture medium contained high concentrations of ferrous iron all strains appeared to influence corrosion to the same degree, corrosion/time curves were linear, and there was no correlation again between corrosion rates and hydrogenase activity. The corrosion rates were high: values of the order of 220 mg dm^{-1} day^{-1} were obtained. Under these conditions no real film formation occurred on the metal, but instead a bulky black cocoon of corrosion product formed around the metal specimen. An interesting feature of this corrosion product is that it consisted of an equimolar mixture of ferrous sulphide and ferrous carbonate (siderite, $FeCO_3$). This has been discussed elsewhere[19].

Semicontinuous cultures in high-iron medium (0·5% crystalline ferrous ammonium sulphate)[20] gave broadly similar results to the high-iron continuous experiments.

At about the same time it was discovered that vigorous depolarization of cathodes was effected by ferrous sulphide itself, and that once such a bulky layer had formed in semicontinuous culture in high-iron medium its depolarizing activity continued for a long period after change-over to a low-iron medium[21]. In addition it has been shown that chemically prepared ferrous sulphide can act as a depolarizing agent in polarization experiments in the absence of bacteria[22], and stimulates the uptake of molecular hydrogen for bacterial sulphate reduction in manometric experiments[23]. That ferrous sulphide plays an important role in anaerobic corrosion is therefore certain.

Mara[24] recently found, using a continuous culture technique in minimal-iron medium, that the time for breakdown of the protective sulphide film to occur was dependent on the growth rate of the bacteria (governed by the dilution rate), as was the structure of the film. After film breakdown, the corrosion rate was high and apparently not dependent on growth rate. In high-iron medium, corrosion rate was independent of growth rate except when the latter was very low.

To attempt to establish that hydrogenase is an important contributory factor in the corrosion process entailed avoiding the complications due to the presence of sulphide ion. Utilizing the discovery that certain strains of the sulphate-reducing bacteria can grow on fumarate in sulphate-free medium[25], it was shown that actively growing cultures of hydrogenase-positive strains did indeed cause depolarization, although in this system the corrosion rates for mild steel were very low[22]. In this connexion, it is interesting to note that several workers have tried to determine whether hydrogenase-positive organisms other than the sulphate-reducing bacteria can contribute towards anaerobic corrosion by causing depolarization. Findings have been inconsistent: Booth, Elford and Wakerley[26] concluded that neither hydrogen bacteria nor methane bacteria participate significantly in this process, while Mara[24] recorded active depolarization by suspensions of certain photosynthetic and non-photosynthetic microbes.

The available evidence, then, suggests that the factors controlling anaerobic microbiological corrosion are (*a*) the utilization of hydrogen by sulphate-reducing bacteria (and possibly by other microbes that possess a suitable enzyme system); (*b*) cathodic depolarization by precipitated ferrous sulphide; (*c*) the prevention of formation of protective sulphide films in the presence of excess ferrous ions; (*d*) anodic stimulation by the sulphide ions; and perhaps (*e*) the formation of local concentration cells. It would seem that (*a*) and (*b*) are of most importance, and are perhaps interrelated to some extent in a way that is not yet fully understood. Factor (*b*) is dependent upon the concentration of ferrous ions in the system, and so this latter may well have some bearing on the corrosion behaviour of buried metals and could therefore be a significant factor in determining the potential aggressiveness of a soil. Indeed the data presented in Tables 3.3 and 3.4 indicate that corrosion occurred in a soil having a mean ferrous ion content over a year of 333 μg/g soil, but not in a soil of mean content 59 μg/g.

Corrosion in Natural Environments

Corrosion in the Soil

As we have already seen, underground corrosion of unprotected ferrous metal objects is often most severe in wet clay or clayey soils of about neutral pH. Progressive pitting corrosion is common: cast-iron pipes of wall thickness $\frac{1}{4}$ inch (0·63 cm) have occasionally become perforated within a year of installation in these conditions, while perforation within four years is quite common. Such corroded objects, if examined immediately on removal from the soil, have a black corrosion product frequently smelling of H_2S (or which liberates H_2S on treatment with acid: this should be used on the site as a test to distinguish the corrosion product from black magnetite). The sulphide-containing corrosion product is often loose, and when lifted reveals pits lined with bright metal (the anodic areas). A type of corrosion peculiar to cast iron is *graphitization*: in certain areas the iron has been leached out leaving a matrix of graphite, often indistinguishable from the rest of the iron pipe except on closer inspection, but soft enough to be pierced by sharp instruments or to give way under pressure. Typical examples of pipes that have corroded in the sort of conditions just described are shown in Figs. 3.10 and 3.11.

Fig. 3.10 *Cast iron pipe section turned to show pitting by bacterial corrosion (Crown Copyright reserved)*

When these symptoms are discovered in such environments, it can be assumed that the sulphate-reducing bacteria are probably responsible; and they can in fact usually be demonstrated in the corrosion product and the soil in the vicinity of the pipe in very much larger numbers than the average "background" count for that district. In other words, the presence of a ferrous object would appear to have rendered the local conditions more favourable for the growth of sulphate reducers, perhaps by providing a supply of cathodic hydrogen or of ferrous ions which render the redox potential more negative. Once the bacterial growth has received a slight stimulus in this way, it probably becomes a self-reinforcing process: a little sulphide ion will be produced which in turn further depresses the redox potential and stimulates more growth. This can only be regarded as conjectural, though, until more studies are made.

There have been several attempts to pinpoint the physical and chemical

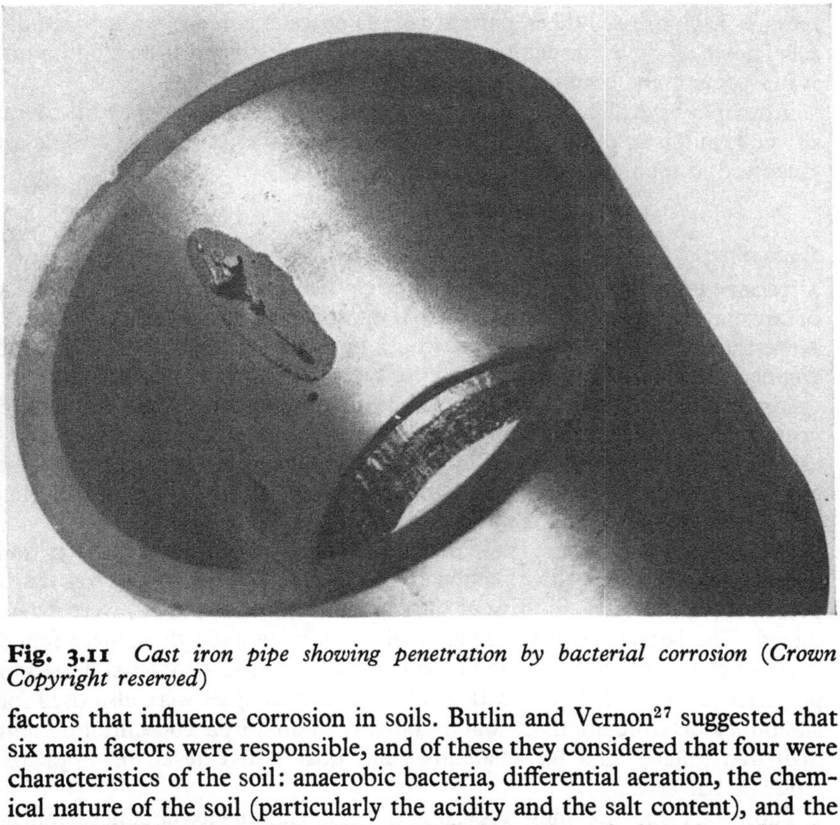

Fig. 3.11 *Cast iron pipe showing penetration by bacterial corrosion (Crown Copyright reserved)*

factors that influence corrosion in soils. Butlin and Vernon[27] suggested that six main factors were responsible, and of these they considered that four were characteristics of the soil: anaerobic bacteria, differential aeration, the chemical nature of the soil (particularly the acidity and the salt content), and the formation of concentration cells. A different point of view was taken by Starkey and Wight[28], who considered that the activity of sulphate-reducing bacteria ought to reflect the redox potential of the soil. From their studies they compiled the following table:

Soil E_h (redox potential on hydrogen scale)	Corrosivity
$<$100 mV	severe
100–200 mV	moderate
200–400 mV	slight
$>$400 mV	non-corrosive

Romanoff[29], in a useful review of underground corrosion in 1957, regarded the important factors as being aeration, the nature of the electrolyte, electrical factors such as the number and arrangement of anodic areas, and miscellaneous in which he included the presence of sulphate reducers. According to Romanoff, soil resistivity correlates well with the corrosiveness of soils; other workers have stressed the importance of pH or water content, while Stratfull[30] has shown a close correlation between soil aggressiveness, an "optimum" water content and a minimum resistivity.

Attempts have also been made to correlate soil aggressiveness with counts of viable sulphate-reducing bacteria. The relative merits of these criteria are discussed in the next section.

Evaluation of Criteria of Soil Aggressiveness

It became clear that insufficient data were available to establish the reliability of any one, or any combination, of the proposed criteria of soil aggressiveness. Nevertheless such reliable criteria would be extremely useful to a corrosion engineer, and therefore Booth and colleagues[31,32] devised a large-scale investigation from which it was hoped to determine the validity of various criteria. This study entailed periodic measurements of a number of factors at 87 sites distributed over England and Wales. The corrosivity of 59 of these sites was already documented by local authorities, gas and water boards etc., and of these 59 sites, 41 were known to be aggressive.

The following tests were made periodically over five years. At the sites, soil resistivity (indicative of the likelihood of oxidative corrosion) and soil redox potential (considered indicative of the risk of bacterial corrosion) were determined. The latter of course required pH determinations; these were found to be more reliable when made in the laboratory on soil samples collected in polystyrene jars with tightly fitting lids. The soil samples were also used for laboratory determinations of water content, soluble iron concentration and hydrogen uptake, and for qualitative and quantitative tests for sulphate-reducing bacteria.

Tables 3.3 and 3.4 show typical sets of measurements obtained from a corrosive and a non-corrosive site respectively; and Figs. 3.12 and 3.13 show resistivity and redox potential readings taken during the survey and plotted as histograms.

Consideration of all the data in relation to the history of the sites, when known, suggested that the aggressive sites were characterized by a mean soil resistivity of less than 2000 ohm.cm or a mean redox potential more negative than +400 mV with respect to the normal hydrogen electrode when corrected to pH 7 (+430 mV in the case of a predominantly clay soil). These criteria would predict that 38 sites were aggressive out of the 41 that were actually known to be. They were also used to predict that six of the 28 sites of unknown aggressiveness would prove aggressive; this prediction was subsequently confirmed.

It is interesting to note that in the two histograms (Figs. 3.12 and 3.13), in which the aggressive sites are subdivided according to whether the aggressiveness was attributed to low resistivity or to low redox potentials, the critical resistivity value of 2000 ohm.cm agrees exactly with the value below which

Table 3.3 *Results of tests on a soil known to be aggressive (Crown Copyright reserved)*

Name of site: Haverfordwest.
Description of soil: Heavy, black, waterlogged and marshy soil.
Corrosion history: Corrosion of pipelines recorded by Wales Gas Board

	Resistivity Ω-cm at 3 ft	Redox-potential, V at pH = 7 (N.H.E.)	Water content, %	'Bottle test', days to blacken		Viable count of sulphate-reducers, cells/g	H_2 uptake, ml	pH	Fe, μg/g
				Baars' medium	Baars' medium + cysteine				
January	2,400	0·060	53	9	3	10^5	—	6·0	420
February	3,260	0·040	66	9	3	10^4	—	6·0	1,000
March	1,630	0·040	59	9	2	10^6	—	6·1	104
April	2,300	0·070	52	5	6	10^4	—	5·9	250
May	2,010	0·132	61	4	4	10^4	—	4·3	420
June	1,680	0·065	49	13	4	10^5	41	5·5	166
July	1,150	0·150	46	6	3	10^6	—	6·5	86
August	1,680	0·070	51	11	4	10^5	55	6·0	325
September	2,400	0·070	44	7	2	10^4	—	6·2	390
October	1,150	0·160	49	8	3	10^7	40	6·7	42
November	2,300	0·100	54	7	3	10^6	—	6·6	120
December	3,350	0·070	53	12	4	10^3	51	6·0	720
Mean	2,110	0·064	53	8	4	10^5	47	5·9	333

Table 3.4 *Results of tests on a soil known to be non-aggressive (Crown Copyright reserved)*

Name of site: Heston.
Description of soil: Light brown loam with good crumb formation.
Corrosion history: No incidence of corrosion reported by Metropolitan Water Board.

	Resistivity, Ω-cm at 3 ft	Redox-potential, V at pH = 7 (N.H.E.)	Water content, %	'Bottle test', days to blacken: Baars' medium	'Bottle test', days to blacken: Baars' medium + cysteine	Viable count of sulphate-reducers, cells/g	H_2 uptake, ml	pH	Fe, μg/g
January	4,790	0·528	13	6	3	10^5	—	5·8	40
February	9,580	0·350	14	8	4	10^5	—	5·3	40
March	4,790	0·504	18	10	3	10^4	—	5·9	130
April	9,580	0·466	16	neg.	32	10^3	0	5·3	0
May	4,790	0·550	16	7	3	10^4	—	7·4	54
June	7,660	0·530	14	5	4	10^5	—	5·3	50
July	9,580	0·540	10	6	3	10^4	—	6·0	50
August	26,820	0·548	10	8	3	10^4	0	5·7	30
September	8,620	0·510	14	9	2	10^4	25	7·3	66
October	14,370	0·550	16	5	2	10^4	—	5·8	71
November	9,580	0·520	18	6	3	10^4	—	6·5	45
December	4,790	0·560	15	7	2	10^4	0	6·3	134
Mean	9,560	0·513	15	7	5	10^4	6	6·0	59

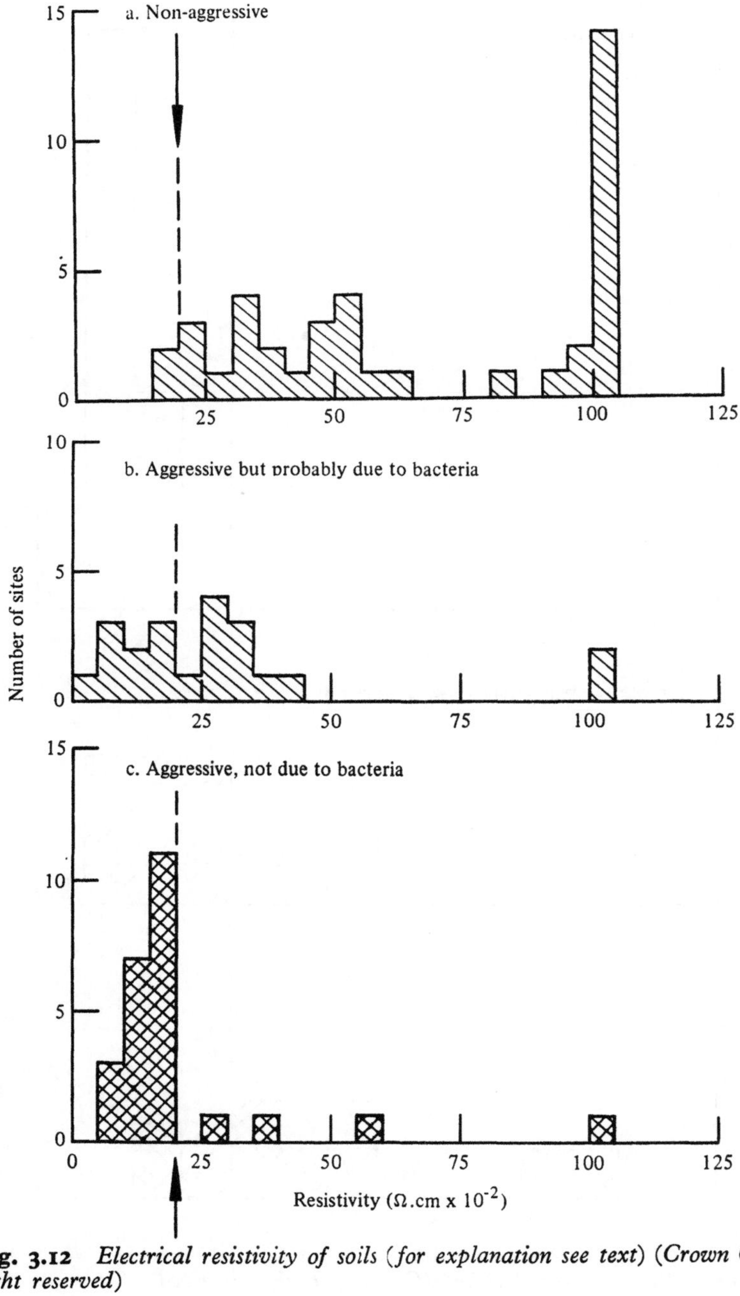

Fig. 3.12 *Electrical resistivity of soils (for explanation see text)* (*Crown Copyright reserved*)

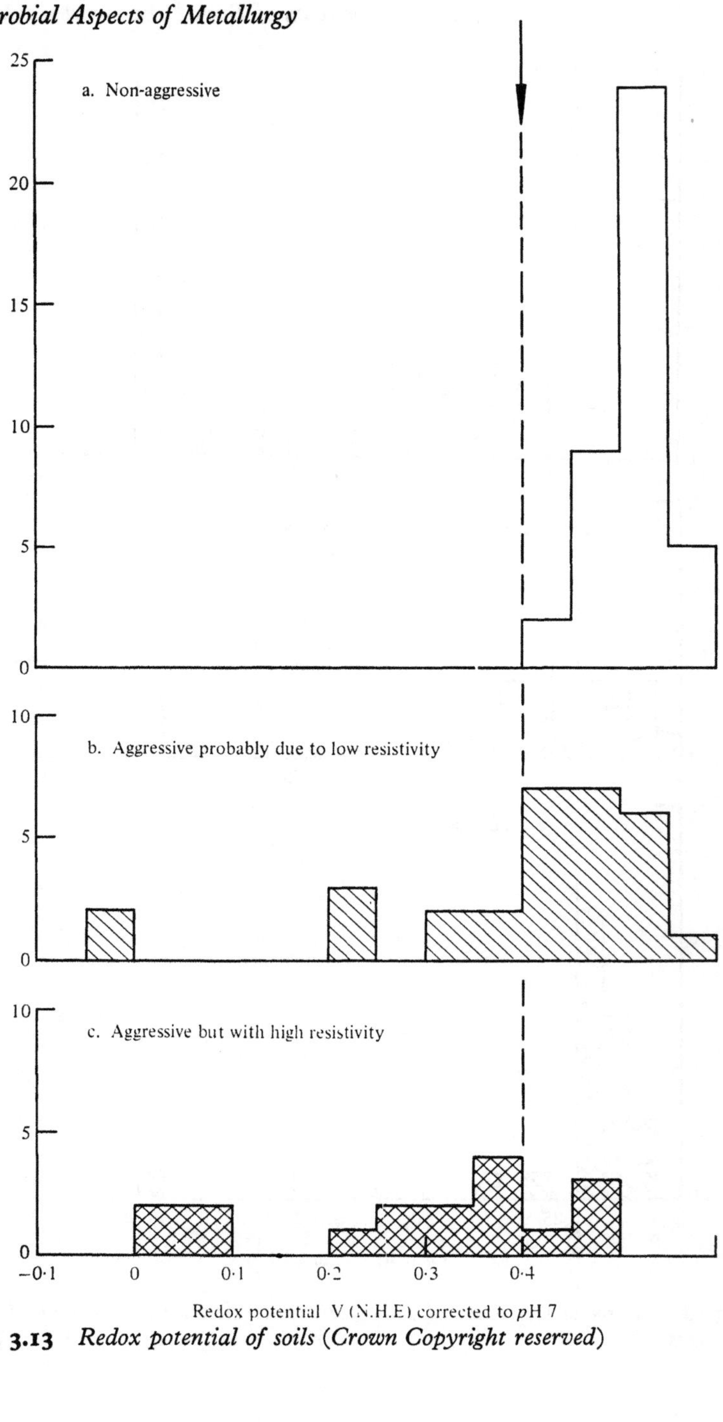

Fig. 3.13 *Redox potential of soils (Crown Copyright reserved)*

Schwerdtfeger[33] recommended protection, and that the value of 430 mV for critical redox potential is very close to the figure for non-aggressiveness given by Starkey and Wight[28] and quoted above.

When a soil was considered a "borderline" case because its resistivity was close to 2000 ohm.cm or its redox potential was within the range 400–430 mV, it was classified according to the water content: if this was greater than 20% the soil was regarded as aggressive, and if below this figure, non-aggressive. One borderline case, however, was not correctly resolved by this method, and it was considered that other factors were operating at this site which could not be detected.

Four sites did not conform at all. Of these, two were assessed as aggressive; since no industrial pipework was buried at one of them the prediction could not be confirmed, while the only pipework buried at the second site was situated on an embankment in made-up ground (a non-indigenous substratum composed of a mixture of soil and clay excavated from other sites in the district plus various kinds of rubbish, and often highly corrosive). The remaining two sites had been predicted as non-aggressive, but the local authorities reported several cases of corrosion occurring in buried pipelines; this was attributed to the fact that most of the breakdowns occurred on a housing estate where a great deal of made-up ground was interspersed with the virgin soil.

As regards hydrogen uptake by the soils, higher values were found in the more aggressive soils. Similarly, all soils with a mean soluble iron content greater than 120 μg/g were aggressive. Interesting though these results were because of their fulfilment of predictions from laboratory work, the estimations themselves were time-consuming and their use in diagnosing aggressive soils is not recommended.

Sulphate-reducing bacteria were present in all the 87 soils studied. The quantitative tests showed that the viable count was higher – often by several orders of magnitude – in soils of low redox potential. Since, however, it is much simpler to make an E_h measurement (though no satisfactory probes are available, and some practice is required with home-made ones constructed to the specifications in reference (31) to obtain meaningful results) than to carry out bacterial tests, the E_h measurement is recommended. In cases where resistivity and E_h measurements cannot be made *in situ* for various reasons, however, other methods such as bacteriological growth tests may have to be resorted to.

The reliability of the prognostic tests was confirmed by a programme of work[34] in which previously weighed plates of mild steel, aluminium, copper, zinc and lead were buried at 12 sites that had already had their aggressiveness predicted by resistivity, redox potential and water content measurements. Periodic measurements of metal-to-soil potential (with reference to a copper/copper sulphate electrode) and redox potential of both the bulk of the soil and of that within a few inches of each specimen were made, and the specimens were exhumed after 1, 3 or 4 years for weight-loss determination.

The predictions based on the criteria previously recommended[32] were found to be valid for mild steel except in two cases where previously unnoticed local conditions had caused more corrosion than was anticipated. The method was also shown to be fairly satisfactory for lead and zinc, but of relatively little use for predicting aluminium corrosion. None of the soils was aggressive towards copper. The metal-to-soil potential measurements gave little information of value.

The relationship between soil type and corrosion was brought out clearly in these investigations, as is shown in Table 3.5.

Table 3.5 *Weight loss of metal plates during a year's burial in various types of soil (Crown Copyright reserved)*

Type of soil	Weight loss ($mg/dm^2/day$)				
	Mild steel	Al	Cu	Zn	Pb
Clay	9·7	0·1	0·7	13·1	1·2
Clay-loam	11·3	0·4	0·9	2·8	4·1
Loam	5·6	0	0·3	3·3	0·4
Chalk	5·3	0·03	0·3	7·7	0·3

Corrosion in Rivers

The tidal waters of the River Thames were tested for corrosiveness by exposure of pre-weighed mild steel plates for periods of up to five years, and attempts made to correlate corrosiveness with physical, chemical and biological factors[35,36,37]. The plates were mounted in such a way that they were buried in the bottom mud, permanently immersed or half immersed (by being supported in floating frames) or alternately immersed and exposed to the air by tidal action. When the plates were removed, weight loss was determined and the corrosion product analysed. Water samples were taken for analysis and for counts of sulphate-reducing bacteria.

Plates exposed for one or two years showed both aerobic and anaerobic (microbiological) corrosion. Maximum weight loss and depth of pitting occurred in those plates that had been immersed in the most polluted part of the river. X-ray diffraction patterns of the corrosion product revealed the presence of α-$Fe_2O_3.H_2O$, γ-$Fe_2O_3.H_2O$ and Fe_3O_4. FeS and $FeSO_4$ were also sometimes detectable. That they were not found more often was attributed to their complete removal after oxidation during the cyclical shifts to more oxygenated conditions in the river.

These coatings of corrosion product appeared to confer protection upon the metal specimens: the rate of weight loss gradually decreased over the next three years. During that period the river also became less polluted, and the salinity of the estuarine water then proved to be the most important single factor governing corrosion.

The corrosion product on metal plates from the bottom mud was, in all cases but two, a mixture of FeS, $FeSO_4$ and Fe_3O_4, indicating that sulphate-reducing bacteria are highly active in such situations. The bacteria themselves were demonstrated in all the water samples taken, though the count ranged from about 10 organisms per ml at the nearest point to the open sea from which samples were taken (where pollution was low and dissolved oxygen concentration high) to about 10^5 per ml in highly polluted reaches of the river. In addition, sulphate reducers were demonstrated in the corrosion product from all plates.

It will be seen, then, that the corrosion picture is very complex in an environment such as the lower reaches of a river with its variation in salinity and level of pollution and seasonal variation in oxygen content. Nevertheless there is clearly a significant contribution by the sulphate-reducing bacteria, particularly in the bottom mud, while even in aerated waters there is evidence that sulphate reducers participate in corrosion, presumably by finding favourable micro-habitats in the corrosion product already present.

The Prevention of Corrosion of Buried and Immersed Metals

Introduction

There are various approaches to the problem of preventing underground and underwater corrosion, though all suffer from certain practical disadvantages so that there is no panacea. The following scheme for classifying these approaches is a modification of that of Hoar[38]; it was devised with underground installations in mind, but can be applied to aqueous situations where appropriate.

Methods of preventing or retarding underground corrosion:

1. Using non-corrodible materials.
2. Using corrodible materials
   - (*a*) in a non-aggressive environment
   - (*b*) in an aggressive environment
     - (i) giving the installation a non-aggressive surround
       1. by arranging that the environment contains nothing that will accept electrons or cations.
       2. by using a biocide in the environment to prevent the growth of sulphate-reducing bacteria.
     - (ii) making the metal sufficiently negative with respect to its environment to prevent cations escaping into it.

(iii) arranging a barrier, impervious to cations and/or electrons, between the metal and its environment
1. by application of protective coatings.
2. by the spontaneous or controlled development of conversion coatings or passive films on the metal.
3. by the spontaneous deposition on to the metal of various protective substances such as calcareous scale.

Most of these methods are dealt with in more detail below.

Use of Non-corrodible Materials (Method 1 *in the above scheme)*

Clearly the problem of underground corrosion could be avoided altogether by the use of chemically inert materials of adequate strength. High-silicon iron and austenitic chromium-nickel steels show high chemical resistance, but the former is too brittle for most purposes while the austenitic steels are very expensive. Copper also falls in the latter category. (Strictly speaking, stainless steel comes into category 2(b)(iii)2). Non-metallic materials that have proved satisfactory for pipes despite their low tensile strength include asbestos cement and plastics such as PVC and polyethylene. The plastics used should be free from plasticizers, since there is evidence that these can be microbially degraded.

Concrete is quite often used as an expensive alternative to steel for piles (a report by Romanoff[39] that steel piles suffer much less corrosion than pipelines should be treated with caution in view of catastrophic corrosion of piles in aggressive soil in Australia); but concrete suffers from attack in acid soils, in high sulphate ion concentration, and sometimes through the action of the sulphur-oxidizing bacteria.

No material can approach the tensile strength of mild steel on a cost basis, and hence it is likely to remain the most used material in the foreseeable future for the structures we are considering.

Construction of a Non-aggressive Surround (Method 2(*b*)(i)1)

Theoretically, Method 2(*b*)(i)1 demands that (*a*) no electron acceptors (the principal ones being molecular oxygen and hydrogen ions) should be present, and (*b*) water should be absent, to prevent cations forming from anodic metal. These conditions are obviously impossible to achieve in natural environments; and in fact, as we have already seen, the absence of oxygen will, if sulphate-reducing bacteria are present, itself result in severe corrosion. A fruitful approach is to aim for alkaline conditions and for good drainage – the latter to ensure oxygenation, and hence suppression of the activity of sulphate reducers, rather than to attempt to reduce anodic dissolution.

Such a non-aggressive surround for a pipeline can be made by completely surrounding it with at least 25 cm of chalk. Sometimes a chalk/sand mixture is used for cheapness; or alkalinity may be dispensed with altogether by using gravel, rubble or clay-free sand. In all cases gradients must be controlled and soakaways provided at low points.

Use of Biocides (*Method* 2(*b*)(i)2)

In general, it would not be expected that a biocide, applied locally around a pipe before backfilling, would persist for long enough to be effective against sulphate-reducing bacteria for more than a fraction of the desired life-span of the pipeline, unless the substance were only slightly soluble and yet bacteriostatic at that low concentration even in the presence of organic matter etc. Nevertheless, a study reported in Chapter 4 under *Pipeline Corrosion* indicates that microbial corrosion could be inhibited by experimental treatment with a biocide. More studies could profitably be made on this aspect of corrosion control.

In closed (or relatively closed) industrial systems, of course, the control of sulphate-reducing bacteria with biocides is easier, though as will be seen in Chapter 4 the programme of treatment in these complex environments with mixed microfloras usually has to be determined empirically.

Cathodic Protection (*Method* 2(*b*)(ii))

The principle of this method is to make the metal-to-environment potential of the structure to be protected so negative that cations cannot pass into solution from the metal. The structure is therefore made cathodic with respect to another electrode installed for the purpose in the neighbourhood. Two methods are used. Sacrificial anodes of a baser (more electronegative) metal than iron – magnesium, aluminium or zinc – may be employed; the relatively low p.d. that can be obtained limits the effective range of a sacrificial anode, but there is the compensating advantage of minimal interference with other installations. The alternative method is to construct a large anode or ground-bed, often of graphite, fairly remote from the structure to be protected, and apply an e.m.f. (for example, from the mains *via* a rectifier) between the two. Theoretically a carbon ground-bed should be everlasting, but in practice occasional renewal is needed.

The choice between installation of sacrificial anodes and the use of an impressed e.m.f., then, usually depends on local conditions such as soil resistivity and availability of a source of cheap electric power. The correct application of cathodic protection requires an accurate knowledge of the potential of the structure relative to the corroding medium (soil or water) both before and after the application of the current. The standard way of measuring pipe potentials is by the copper/saturated copper sulphate electrode; and the standard practice in cathodic protection is to position the anodes or adjust the resistance of the circuits so as to depress the potential of steel to about −0·85 V relative to that electrode, equivalent to −0·53 V (NHE).

Cathodic protection can also prevent corrosion by the sulphate-reducing bacteria, but here a more negative potential is required. Horváth and Novák[40] calculated the requirement in presence of sulphide; a value of −0·95 V ($Cu/CuSO_4$ electrode) is usually accepted, though these authors showed that in some cases an even more negative potential may be required.

The alkalinity of the catholyte probably assists by suppressing the growth of sulphate reducers; but on the other hand over-protection will result in over-alkalinity which in turn can damage certain types of protective coatings (of which more below), causing blistering and peeling.

Investigations of the cathodic characteristics of mild steel in the presence and absence of sulphate-reducing bacteria by Booth and Tiller[41] have provided evidence in favour of the existing practical criterion for cathodic protection in situations with bacterial activity. The relationships between hydrogenase activity, quantity of cathodic hydrogen liberated and utilized by the bacteria, and the amount of iron corroded at the anode, were studied in the following way. A mild steel electrode was maintained at a constant potential and the current supplied from the potentiostat to maintain the potential, and the loss of iron (determined by analysis of the anolyte) from a counter electrode, were recorded against time. The equivalence between anolyte iron and coulombs drawn from the potentiostat was determined.

Using washed non-growing suspensions of *Dv. vulgaris* strain Hildenborough (hydrogenase-positive) and *Dt. orientis* (hydrogenase-negative) in an electrochemically inert buffer of high conductivity with benzyl viologen as electron acceptor (thus avoiding complications due to growth, concomitant depletion of medium, and sulphide ion and FeS formation), the rate of change in potential due to reduction of the dye to reach the mid-point potential was recorded.

In the absence of the reducible substrate, the depolarization of the cathodes was independent of the hydrogenase activity of the bacteria whereas in its presence the cathodic depolarization was proportional to the hydrogenase activity. The rate of reduction of the substrate was also proportional to the hydrogenase activity. As an example, the curves obtained at a working potential of −0·85 V (NHE) are shown in Fig. 3.14. These results provide evidence in favour of the theory[5] that sulphate-reducing bacteria act as depolarizing agents.

If the logarithm of the extra current required to keep the cathode polarized in the presence of the bacteria over that required in buffer alone (log Δi) is plotted against enzymic activity of the organisms at each controlled potential, a straight-line relationship is found (Fig. 3.15) similar to that of a Tafel plot; this indicates that in the presence of a reducible substrate log Δi is proportional to the hydrogenase activity ($-Q_{H_2}^{BV}$). This is particularly striking at the lower controlled potentials of the electrode; and if the slopes of these lines

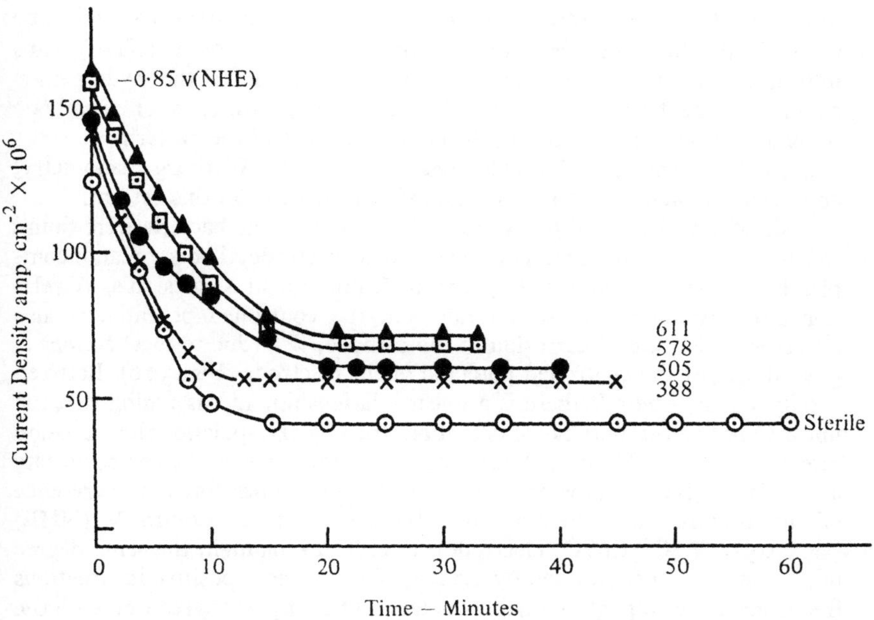

Fig. 3.14 *Current/time curves for* Dv. vulgaris *system with benzyl viologen as electron acceptor, at a controlled potential of −0·85 V. The numbers against the curves represent values of* $-Q_{H_2}^{BV}$ (μl/mg/h) *(Crown Copyright reserved)*

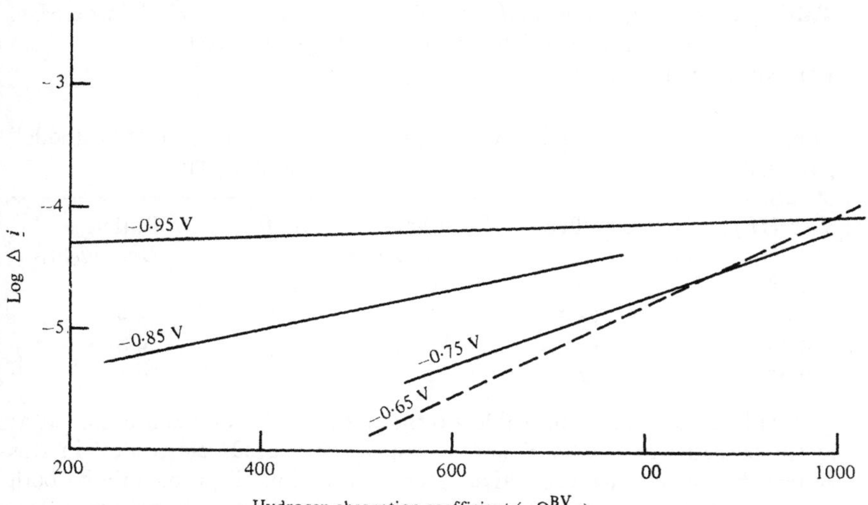

Fig. 3.15 *Relationship between 'bacterial current' and hydrogenase activity for* Dv. vulgaris *at various controlled potentials (Crown Copyright reserved)*

($d(\log \Delta i)/dQ^{BV}_{H_2}$) are plotted against the controlled potentials a straight line is obtained which intercepts the potential axis at −0·98 V (NHE), thus indicating that the extra current Δi is independent of the hydrogenase activity of the bacteria and that the rate of depolarization of a cathode becomes constant irrespective of the enzyme activity of the bacteria.

Dt. orientis, shown to be hydrogenase-negative by Warburg manometry, appeared completely inactive as a cathodic depolarizer in this system.

A significant feature of this work is the effect that the bacteria were found to have on local anodic action on the cathodic electrode, that is, on an incompletely cathodically-protected specimen. Table 3.6 illustrates this. A relationship between the corrosion rate and the controlled potential is also observed if the rate of corrosion of the cathode, W_c, (due to local action) is plotted against the controlled potential of the electrode (Fig. 3.16). Between −0·85 V and −0·65 V there is a linear relationship, and assuming that the linearity persists at more noble potentials, then by extrapolation the corrosion rate at −0·510 V (NHE) in buffer solution is the same as the corrosion rate at −0·618 V (NHE) in the presence of the bacteria. Therefore in the presence of the bacteria the potential must be depressed to −0·618 V (NHE) (= −0·958 V (Cu/$CuSO_4$ electrode)) in order to maintain the same degree of protection as that afforded by a potential 0·1 V more positive in situations free from active sulphate-reducing bacteria. These figures agree well with the generally-accepted values quoted above, and with those produced by other workers such as Whalley[42].

Table 3.6 *Catholyte iron analyses and corrosion rates of "cathode" at various controlled potentials* (*Crown Copyright reserved*)

For explanation see text.

Controlled potential of cathode V (NHE)	Catholyte iron μg/h		Corrosion rate of "cathode" mg/dm²/day	
	Buffer	Buffer + *Dv. vulgaris*	Buffer	Buffer + *Dv. vulgaris*
−0·95	0	0	0	0
−0·85	25·5	33	3·1	3·9
−0·75	52·0	82	6·3	9·8
−0·65	112·5	221	13·5	26·5

It will be observed from Table 3.6 that measurable corrosion occurs at as low a potential as −0·85 V (= −1·19 V (Cu/$CuSO_4$ electrode)) in this system, but such a low protective potential would in all probability be both uneconomic and a possible danger to the structure since hydrogen embrittlement of the metal could occur, while the effects of the alkali produced have already been mentioned.

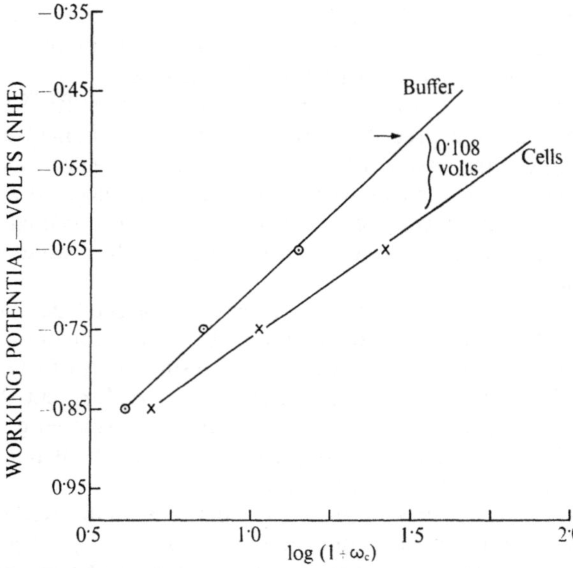

Fig. 3.16 *Relationship between potential and anaerobic corrosion (Crown Copyright reserved)*

The principles of cathodic protection have been dealt with at length by Morgan[43].

Protective Coatings (Method 2(*b*)(iii)1)

The concept of protective coatings for base metals is by no means new, and yet it is still doubtful whether there is an ideal coating substance that is adherent, coherent, completely non-porous, mechanically resistant to the hazards encountered during delivery, laying and backfilling and chemically resistant to prolonged contact with all kinds of natural environments.

Coal tar-based materials and asphaltic bitumen-based materials are frequently used; the former are said to be more stable and waterproof. It has long been suspected that organic reinforcements to these coatings, such as hessian, constitute if anything a weakness, and it was later demonstrated that they could be attacked and broken down by cellulose-decomposing microbes. A much better reinforcing material for bituminous coatings and linings is fibreglass. All joints, valves etc. in such protected pipelines must be coated with bitumen after assembly, and damaged parts of the original coating made good: corrosion is often extremely rapid at small discontinuities or "holidays" in protective coatings.

Concrete linings have sometimes been applied to pipes. Apart from their alkalinity they have rather little to offer in the way of protection, being usually permeable to air, water and stray currents unless they are very thick – and hence expensive. Corrosion of the encased metal is particularly likely to occur at interfaces between different castings of the concrete.

The coating of steel pipes with zinc at a rate of 2·8 oz/ft^2 (854 g/m^2) has been stated to prevent appreciable corrosion for 10 years in neutral or alkaline soil. Zinc coatings are unsuited for acid conditions, however, and must not be used in electrolytic contact with a large area of uncoated steel. Spray-applied zinc-aluminium coatings show promise[38]. Lead-coated pipes have been tested; they performed well for a time, but once the coating failed rather rapid corrosion ensued.

Not surprisingly, plastic coatings have received a good deal of attention. The main problems seem to be the bonding of the plastic material to the pipe and the *in situ* coating of joints. Fibre-glass resin and epoxy resin coatings are stated to be highly protective, though expensive. A recent development dispenses with the property of adhesion, formerly considered essential for a coating: this is the "bagging" technique in which the pipe is put into a loose plastic sleeve. It is stated to be effective.

Logan and Parker[44] reviewed protective coating systems available in 1965.

The difficulty of achieving a completely unbroken, completely non-porous coating is so great – or, looked at from the other aspect, the expense of applying cathodic protection to a large metal installation is so high in terms of power or sacrificial anodes consumed – that for best results in an economic context a combination of these methods is often employed.

Other Types of Microbial Corrosion

Microbial Corrosion due to Acid Production

An economically important group of bacteria, the thiobacilli, are dealt with several times in this volume. They obtain energy from the oxidation of sulphur and its more reduced compounds including H_2S (see Chapter 1, and Chapter 4 under *Combined Action of the Sulphur Bacteria in Metallic Corrosion*), and in addition the species *Thiobacillus ferro-oxidans* can oxidize ferrous iron. The free sulphuric acid that certain species produce (or which, in the case of ferrous iron oxidation, arises as a secondary reaction as explained in Chapter 7) can naturally cause serious corrosion of metals underground: it is stated that they can produce, and remain viable in, solutions of 10–12% H_2SO_4[45].

These organisms are aerobes (with one exception), and in any case of severe corrosion in aerobic conditions, where the local pH is very low, sulphur-oxidizing bacterial activity should at once be suspected. A case recorded in the literature[46] concerns a gas main – actually bitumen-covered –

that corroded in a peaty soil within 2½ years. In the vicinity of the pipe a pH as low as 2·0 was recorded, and sulphuric acid demonstrated. At lower, oxygen-free levels in the soil were active sulphate-reducing bacteria, from which sulphide was diffusing upwards. This was then oxidized to H_2SO_4 by thiobacilli. Drainage and treatment with lime were recommended: thiobacilli will not grow in alkaline conditions.

Corrosion by acids is dealt with in Chapter 2, and will not be elaborated upon here. It is clearly necessary for protective coatings to be acid-resistant.

Organic acids are produced by a large number of microbes, and their effects may sometimes be serious. It has already been mentioned under *Protective Coatings* that cellulose-decomposing microbes damage hessian wraps. The hessian is broken down to give lower fatty acids such as acetic and butyric under anaerobic or partially anaerobic conditions. Pitting of the lead sheaths of buried cables protected with a bitumen and hessian wrap has occasionally occurred by this mechanism; it is known as "phenol corrosion" because it was thought to be due to phenols in the bitumen until its microbial origin was discovered.

Certain moulds produce organic acids in fully aerobic conditions and have been found to be responsible for cases of corrosion of iron, copper and aluminium protected by insulating material that was capable of being used as a growth substrate by the moulds.

Corrosion due to Heavy Microbial Growths and Deposits

Microbes that form large colonial masses or tufts on the substratum on which they grow (for example algae, fungi, protozoa, bacteria) are always potential agents of corrosion, because oxygen gradients and gradients of solutes present

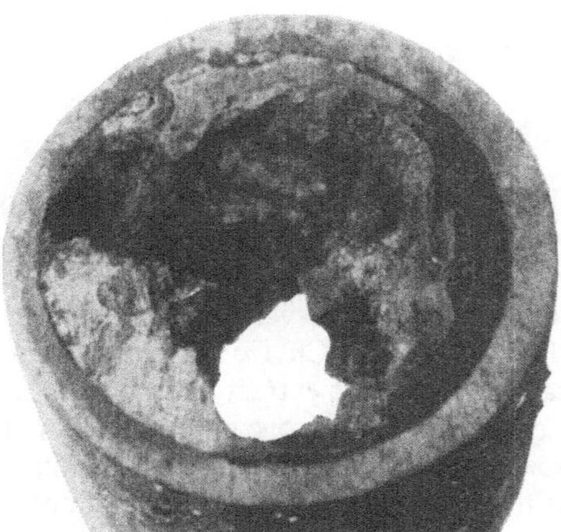

Fig. 3.17. *Advanced stage of tuberculation in a section of cast iron pipe*

in the ambient water and of end-products of the organisms themselves all tend to establish concentration cells. The situation is obviously worse if the organism concerned produces an acid; and the microbiological picture is often very complex as other organisms than the clump-former, such as acid-producers, may become entangled in the growth, or indeed may find it provides good anchorage or a favourable microhabitat. Needless to say, the sulphate-reducing bacteria find suitable conditions in the depths of such a mass of growth, and can cause severe pitting corrosion there.

An example of massive growth is that produced by the iron bacteria, which, together with the ferric hydroxide they produce, can form extensive deposits called tubercles on the inside of iron water pipes. Fig. 3.17 shows an advanced stage of tuberculation in a section of cast iron pipe. Chlorination of the water helps to control tuberculation, confirming its biological origin, but in many cases regular internal cleaning of the pipework has to be resorted to.

APPENDIX

Microbiological Examination of Soil, Water etc. for Sulphate-Reducing Bacteria

Though it may be true that soil aggressiveness is accurately predictable merely from resistivity and redox potential readings, with resolution of border-line cases by determination of water content (and much more work is necessary before this can be finally established), qualitative and quantitative tests for sulphate-reducing bacteria are still an important part of corrosion detection and control. Unfortunately, a fair amount of expertise is necessary, especially for the quantitative tests. The following tests are given for those who have laboratory facilities and a little basic equipment such as an autoclave or pressure-cooker for sterilization.

Qualitative test

The medium in which sulphate-reducing bacteria may most easily be grown is that of Baars, as modified by Postgate[47]. The following is the formulation for 1 litre of this medium (it is best to start with almost a litre of distilled or tap water, adding and dissolving the various reagents, and finally making up to 1 litre): KH_2PO_4, 0·5 g; NH_4Cl, 1·0 g; $CaSO_4$, 1·0 g; $MgSO_4.7H_2O$, 2·0 g; sodium lactate, 3·5 g; yeast extract, 1·0 g; ascorbic acid, 1·0 g; thioglycollic acid, 1·0 g; $FeSO_4.7H_2O$, 0·5 g. Adjust to pH 7·0–7·5. Sterilize immediately after making up, by autoclaving at 15 p.s.i.g. (121 °C) for 15 min. An improvement is to autoclave the appropriate amount of thioglycollic acid separately and add it aseptically to the medium afterwards. This medium should be used immediately after it has cooled: the reducing agents become oxidized fairly rapidly and thereafter the medium no longer has a redox potential negative enough (about $-0{\cdot}1$ V (NHE)) to initiate growth reliably. This applies with even more force to the growth medium for quantitative tests (counts).

Small glass screw-capped bottles of around 10 ml capacity are sterilized, and samples (2 g or 0·2 g) of soil, corrosion product, water etc. placed in them. Sterilized medium is then poured in to fill the bottles completely (the medium has a precipitate, and so should be gently swirled before pouring to mix it thoroughly), and the caps are replaced so as not to trap any air. For examination for *Desulfovibrio* spp., the bottles are incubated at 30° or 37°C; for *Desulfotomaculum nigrificans*, 55°C. In either case blackening of the medium, which usually occurs within 2–3 days, indicates the growth of sulphate reducers. Blackening is due to the formation of ferrous sulphide in this high-iron medium. Note: if the sample is from a marine environment, 2·5% NaCl should be added to this and the following medium.

Quantitative test

This method depends on making serial dilutions of the sample (or of an extract, if it is soil etc.). This is done by preparing and sterilizing a number of cotton-wool plugged test tubes each containing 9 ml of saline (0·5% NaCl in water, or 2·5% for marine samples) or of the medium below with the agar, sodium lactate and yeast extract omitted. Details of this type of microbiological technique, and of the preparation of sterile pipettes, are given in the *Appendix* to Chapter 5.

1 g ($+0{\cdot}01$ g because quantitative transfer is impossible) of the soil sample is ground in a sterilized mortar and 9 ml sterile saline gently "ground in" to give a fine suspension. This is the first dilution (1/10) of the original sample. 1 ml of this is taken up in a sterile pipette and run into a tube of 9 ml saline; this is labelled "1/100". The serial dilution is repeated down to say 1/1 000 000.

The medium to be used has the following formulation: KH_2PO_4, 0·5 g; NH_4Cl, 1·0 g; Na_2SO_4, 1·0 g; $CaCl_2.6H_2O$, 1·0 g; $MgSO_4.7H_2O$, 2·0 g; sodium lactate, 3·5 g; yeast extract, 1·0 g; ascorbic acid, 0·1 g; thioglycollic acid, 0·1 g; $FeSO_4.7H_2O$, 0·5 g; agar, 15 g. Distilled water to 1 litre. Adjust pH to 7·6 with NaOH. Autoclave as above; cool to about 42–43°C.

Aseptically pipette 1-ml portions of the 1/100 dilution into each of 3 or 5 sterile test tubes and label appropriately; repeat for each of the higher dilutions. Pour about 9–10 ml of the cooled (but still liquid) medium into each test tube. The pouring process should mix the saline and the nutrient medium fairly uniformly but a further mixing can be given if desired, by rolling the tube between the hands before the medium sets. After setting has occurred, another 1·5 cm depth of medium is added to each tube to reduce access of oxygen to the inoculated medium below. The tubes can then be incubated as described above.

Black colonies develop in 3–16 days. These are counted for each tube, if not too close together to render this impossible, and the most probable number (MPN) read from tables[48].

References

[1]BOOTH, G. H. '*Sulphur bacteria in relation to corrosion*'. J. Applied Bacteriol., **27**, 174–181 (1964).

[2]VERNON, W. H. J. '*Metallic corrosion and conservation*'. In *The Conservation of Natural Resources*, 105–133. London: Institution of Civil Engineers, 1957.

[3]GAINES, R. H. '*Bacterial activity as a corrosive influence in the soil*'. Industrial & Engineering Chem., **2**, 128–130 (1910).

[4]BENGOUGH, G. D. AND MAY, R. '*Seventh report to the Corrosion Research Committee of the Institute of Metals*'. J. Inst. Metals, **32**, 81–269 (1924).

[5]WOLZOGEN KÜHR, C. A. H. VON AND VAN DER VLUGT, L. S. '*De grafiteering van gietijzer als electrobiochemisch proces in anaerobe gronden*'. Water, Den Haag, **18** (16), 147–165 (1934).

[6]STEPHENSON, M. AND STICKLAND, L. H. '*Hydrogenase. II. The reduction of sulphate to sulphide by molecular hydrogen*'. Biochem. J., **25**, 215–220 (1931).

[7]POSTGATE, J. R. '*The reduction of sulphur compounds by* Desulphovibrio desulphuricans'. J. General Microbiol., **5**, 725–738 (1951).

[8]POSTGATE, J. R. '*Recent advances in the study of the sulfate-reducing bacteria*'. Bacteriol. Revs., **29**, 425–441 (1965).

[9]WANKLYN, J. N. AND SPRUIT, C. J. P. '*Influence of sulphate-reducing bacteria on the corrosion potential of iron*'. Nature, **169**, 928–929 (1952).

[10]WORMWELL, F. AND FARRER, T. W. '*Electrochemical studies of anaerobic corrosion in presence of sulphate-reducing bacteria*'. Chemistry & Industry, 108–109 (1952).

[11]HOAR, T. P. AND FARRER, T. W. '*The anodic characteristics of mild steel in dilute aqueous soil electrolytes*'. Corrosion Sci., **1**, 49–61 (1961).

[12]HORVÁTH, J. AND SOLTI, M. '*Beitrag zum Mechanismus der anaeroben mikrobiologischen Korrosion der Metalle im Boden*'. Werkstoffe und Korrosion, **10**, 624–630 (1959).

[13]BOOTH, G. H. AND TILLER, A. K. '*Polarization studies of mild steel in cultures of sulphate-reducing bacteria*'. Transactions Faraday Soc., **56**, 1689–1696 (1960).

[14]BOOTH, G. H. AND WORMWELL, F. '*Corrosion of mild steel by sulphate-reducing bacteria. Effect of different strains of organisms*'. In *First International Congress on Metallic Corrosion, London, 1961*, 341–344. London: Butterworth, 1962.

[15]TILLER, A. K. AND BOOTH, G. H. '*Polarization studies of mild steel in cultures of sulphate-reducing bacteria. Part 2: thermophilic organisms*'. Transactions Faraday Soc., **58**, 110–115 (1962).

[16]BOOTH, G. H. AND TILLER, A. K. '*Polarization studies of mild steel in cultures of sulphate-reducing bacteria. Part 3: halophilic organisms*'. Transactions Faraday Soc., **58**, 2510–2516 (1962).

[17]BOOTH, G. H., SHINN, P. M. AND WAKERLEY, D. S. '*The influence of various strains of actively growing sulphate-reducing bacteria on the anaerobic corrosion of mild steel*'. In *Comptes rendus du Congres International de la Corrosion Marine et des Salissures, Cannes, 1964*, 363–371. Paris: C.R.E.O., 1965.

[18]BOOTH, G. H., COOPER, A. W. AND COOPER, P. M. '*Rates of microbial corrosion in continuous culture*'. Chemistry & Industry, **86**, 2084–2085 (1967).

[19]BOOTH, G. H. AND TILLER, A. K. '*Some special features of the anaerobic corrosion of steel by organisms of estaurine origin*'. In *Second International Congress on Marine Corrosion and Fouling, Athens*, 1968, 1–3.

[20]BOOTH, G. H., COOPER, P. M. AND WAKERLEY, D. S. '*Corrosion of mild steel by actively growing cultures of sulphate-reducing bacteria. The influence of ferrous iron*'. Brit. Corrosion J., **1**, 345–349 (1966).

[21]BOOTH, G. H., ROBB, J. A. AND WAKERLEY, D. S. '*The influence of ferrous ions on the anaerobic corrosion of mild steel by actively growing cultures of sulphate-reducing bacteria*'. In *Third International Congress of Metallic Corrosion, Moscow, 1967*, vol. 2, section 5, 542–544.

[22]BOOTH, G. H., ELFORD, L. AND WAKERLEY, D. S. '*Corrosion of mild steel by sulphate-reducing bacteria: an alternative mechanism*'. Brit. Corrosion J., **3**, 242–245 (1968).

[23]KING, R. A. AND MILLER, J. D. A. Unpublished results.

[24]MARA, D. D. '*On the anaerobic corrosion of steel by microbes*'. Ph.D. thesis, University of Dundee, 1970.

[25]MILLER, J. D. A. AND WAKERLEY, D. S. '*Growth of sulphate-reducing bacteria by fumarate dismutation*'. J. General Microbiol., **43**, 101–107 (1966).

[26]BOOTH, G. H., ELFORD, L. AND WAKERLEY, D. S. '*Polarization of mild steel in the presence of hydrogen bacteria and methane bacteria*'. Brit. Corrosion J., **3**, 68–69 (1968).

[27]BUTLIN, K. R. AND VERNON, W. H. J. '*Underground corrosion of metals: causes and prevention*'. J. Inst. Water Engineers, **3**, 627–637 (1949).

[28]STARKEY, R. L. AND WIGHT, K. M. *Anaerobic Corrosion of Iron in Soil.* New York: American Gas Association, 1945. 108 pp.

[29]ROMANOFF, M. *Underground Corrosion.* Washington, D.C.: National Bureau of Standards Circular No. 579, 1957. 227 pp.

[30]STRATFULL, R. F. '*A new test for estimating soil corrosivity based on investigation of metal highway culverts*'. Corrosion, **17**, 493t–496t (1961).

[31]BOOTH, G. H., COOPER, A. W., COOPER, P. M. AND WAKERLEY, D. S. '*Criteria of soil aggressiveness towards buried metals. I. Experimental methods*'. Brit. Corrosion J., **2**, 104–108 (1967).

[32]BOOTH, G. H., COOPER, A. W. AND COOPER, P. M. '*Criteria of soil aggressiveness towards buried metals. II. Assessment of various soils*'. Brit. Corrosion J., **2**, 109–115 (1967).

[33]SCHWERDTFEGER, W. J. '*Soil resistivity as related to underground corrosion and cathodic protection*'. National Bureau of Standards J. of Research, **69C**, 71–77 (1965).

[34]BOOTH, G. H., COOPER, A. W. AND TILLER, A. K. '*Criteria of soil aggressiveness towards buried metals. III. Verification of predicted behaviour of selected soils*'. Brit. Corrosion J., **2**, 116–118 (1967).

[35]BOOTH, G. H., COOPER, A. W. AND TILLER, A. K. '*Corrosion of mild steel in the tidal waters of the Thames estuary. I. Results of six months' and one year's immersion*'. J. Applied Chem., **13**, 211–220 (1963).

[36]BOOTH, G. H., COOPER, A. W. AND TILLER, A. K. '*Corrosion of mild steel in the tidal waters of the Thames estuary. II. Results of two years' immersion*'. J. Applied Chem., **15**, 250–256 (1965).

[37]BOOTH, G. H., COOPER, A. W. AND TILLER, A. K. '*Corrosion of mild steel in the tidal waters of the Thames estuary. III. Results of three and five years' immersion*'. Brit. Corrosion J., **2**, 222–228 (1967).

[38]HOAR, T. P. '*Corrosion principles in pipe and pipeline protection*'. In *Proceedings of the 1968 Conference on the Corrosion and Protection of Pipes and Pipelines*, ed. by TRUMAN, W. F. AND MCLOUCHLIN, P., 7–11. London: Brintex Exhibitions Ltd., 1969.

[39]ROMANOFF, M. *Corrosion of Steel Pilings in Soils*. Washington, D.C.: National Bureau of Standards Monograph No. 58, 1962. 22 pp.

[40]HORVÁTH, J. AND NOVÁK, M. '*Potential/pH equilibrium diagrams of some M–S–H_2O ternary systems and their interpretation from the point of view of metallic corrosion*'. Corrosion Sci., **4**, 159–178 (1964).

[41]BOOTH, G. H. AND TILLER, A. K. '*Cathodic characteristics of mild steel in suspensions of sulphate-reducing bacteria*'. Corrosion Sci., **8**, 583–600 (1968).

[42]WHALLEY, W. C. R. '*Cathodic protection and control of corrosion in the Middle East*'. Chemistry & Industry, 140–147 (1954).

[43]MORGAN, J. H. *Cathodic Protection*. London: Leonard Hill Ltd., 1959.

[44]LOGAN, A. G. AND PARKER, W. D. '*Protective Coatings for buried pipes*'. Institution of Gas Engineers J., **5**, 351–366 (1965).

[45]POSTGATE, J. R. '*The microbiology of corrosion*'. In *Corrosion*, vol. 1, ed. by SHREIR, L. L. London: George Newnes Ltd., 1963, 2.51–2.64.

[46]CHEMISTRY RESEARCH 1956. London: Her Majesty's Stationery Office, 1957, p. 16.

[47]POSTGATE, J. R. '*Media for sulphur bacteria*'. Lab. Practice, **15**, 1239–1244 (1966).

[48]MINISTRY OF HEALTH. *The Bacteriological Examination of Water Supplies, 2nd edn.* (Reports on Public Health and Medical Subjects, No. 71). London: His Majesty's Stationery Office, 1939.

CHAPTER FOUR: B. E. PURKISS

Corrosion in Industrial Situations by Mixed Microbial Floras

Microbial Infections in Industrial Waters

As populations explode, and human society becomes more and more dependent on technology, the manufacturing industries seem to grow almost exponentially. It is difficult, if not impossible, to name an industry which has no need for water in its operation; and the ever-growing need for water places an acute strain upon the water resources of most nations so that crisis is never very far away. A slight palliative is the conservation of cooling water by a rigid system of re-use and recirculation. This is justifiable not only on grounds of national need: with water costing about £0·15 per thousand gallons in the United Kingdom and half as much again in Paris, and with prices increasing as water has to be brought to the great industrial centres from ever more remote sources, savings in manufacturing costs can be made despite the capital outlay.

Innovations tend to bring important, and often unexpected, side-effects with them. In a closed or recirculating system,* the side-effect is enhanced microbial infection compared with that in an open-ended, once-through system. This is not, however, surprising when one considers that (1) air-borne microbes of all kinds have easy access to a recirculating system with a cooling tower, just as they have to rivers from which once-through cooling water is taken, (2) a great variety of organic matter providing nutrients for these microbes inevitably finds its way in, and is retained in the recirculated water along with the accumulated microbes, and (3) a wide variety of physical conditions exist that, between them, favour the growth of most microbial types. The cooling tower is the principal offender, providing access for sun-light and ensuring thorough aeration of the water (this latter effect also assists electrochemical corrosion).

The aspect of microbial infection that is our primary concern is the chemical, or metallurgical, effect: corrosion. However, the production of slime is probably just as important economically (indeed, the accumulation of slime contributes to corrosion), while the generation of unpleasant odours can be a significant factor. In all, the cost of microbial infections in Britain's industrial water systems cannot be less than £300 million each year and may be much more. As industry continues to grow, recirculating systems will become the almost inflexible rule; because, in an environment-conscious society, the installation of once-through systems next to rivers, with the attendant changes in the ecological balance downstream from the discharge point, will become unthinkable except in certain very special circumstances. It will thus soon become more essential than ever for the leaders of industry to give due consideration to the problem of microbial infection and enrol the help of the biotechnologist.

* Confusingly, a recirculating system incorporating a cooling tower, and hence open to the atmosphere, is sometimes called an *open* recirculating system to distinguish it from one incorporating a heat exchanger.

There is, then, a new and expanding discipline, Biotechnology, which like so many other recent branches of learning cuts right across the traditional disciplines. The disciplines involved in this case are Biology, Chemistry, Metallurgy and Chemical Engineering; for the biotechnologist is not merely required to identify the cause of corrosion and allied problems (utilizing his biological, chemical and metallurgical skills), but to devise remedies which must be valid on the criteria of cost, facility of application, and side-effects which may conflict with the interests of the community.

In principle, there are three ways of dealing with the problem of microbial growth in industrial situations:

1. Preventing the entry of microbes into these situations. This is virtually impossible.
2. Employing materials that resist the damaging action of microbes. This takes no account of the mechanical effects of microbial growth, described below in some detail under the heading *Recirculating Cooling Systems*; but it is an approach that can be successfully employed, for example, in pumps that handle water in mines containing pyrite (FeS_2). In these environments the bacterium *Ferrobacillus* (*Thiobacillus*) *ferrooxidans* oxidizes the pyritic deposits forming basic ferric sulphate and free sulphuric acid, and lowering the pH to 2 or 3. Such acidic mine waters cause severe damage to pumping machinery and pipes unless these are made of acid-resistent alloy. Chapter 3 deals with the application of this principle to the prevention of underground corrosion.
3. Adding chemicals to the system to kill, or prevent the growth of, microbes. This is usually the only practicable way of dealing with the problem, and will be dealt with at length in this Chapter. The basic principles of chemical disinfection have been dealt with in Chapter 1.

Recirculating Cooling Systems

An industrial recirculating cooling system consists of a heat source or sources (heat exchangers, condensers etc.), a supply of water to which the heat is transferred, and a cooling tower where the heat in the water is given up to the atmosphere. From the microbiological point of view it is most logical to consider the cooling tower first.

The Cooling Tower

This is a device for reducing the temperature of water by passing it in droplet or film form through a current of air. Heat loss to the atmosphere is effected by the evaporation of some of the water, and varies with the temperature of the water (the heat of vaporization of water ranging from 595·5 calories/g. at 0 °C to 539·5 at 100 °C). Obviously, the rate of evaporation, and hence of cooling, will depend on the area of water surface that is exposed to the air current; hence devices to produce a spray of fine droplets or to give a large area of wettable surface are employed. The tower may be designed with a hyperbolic or chimney shape to produce a natural upward convective draught through the falling water; or it may have a fan or fans, either forcing or pulling air up through the tower (forced draught or induced draught towers respectively).

The structural material of the tower may be wood, steel or concrete, and the evaporative surfaces are usually slats of wood, or in some modern structures, plastic materials. The water is introduced into the tower by means of some distributing device which will ensure that the hot water falls evenly over the evaporative surfaces where it can flow down under gravity as a thin film, thus presenting the greatest possible surface area for volume. A common method for achieving this type of distribution is by means of *distribution troughs*, which are wooden troughs with holes bored through the bottom at intervals along their length. These troughs are set in two or three layers: the uppermost or *primary* troughs which deliver the water into the smaller and more numerous *secondary* troughs beneath and at right angles to them, and sometimes a third layer, *tertiary* troughs, at right angles to the secondary troughs. The water passes out of holes in the bottom layer of troughs, often on to ceramic plates which scatter it in drop form, and thence on to the wooden boards or plastic infills down which it drains in film form, ultimately reaching the tower sump or "pond".

A device is incorporated in the system to keep a constant volume of water despite losses due to evaporation, by the addition of *make-up water*. Unless this make-up water is extremely pure, its use leads to a steady increase in the concentration of ions in the cooling water, with an attendant increase in the rate of electrochemical corrosion of the metallic parts. If it contains calcium bicarbonate, scale formation will occur, bringing about a great decrease in heat transfer rate in heat exchangers. When the ionic concentration in the cooling water becomes undesirably high, a portion of it is run off (a process known as *purge*) and the system topped up with make-up water. Purge may be continuous instead of intermittent, when it is called *bleed-off*.

Where the top or sides of the tower are exposed to sunlight there is usually colonization by algae – the "primary producers" of fixed carbon, or organic matter (see Fig. 4.1). These algal growths are often associated with bacteria. They may form in the primary troughs, rapidly reaching the stage where they reduce the bore of the outlet holes, and hence the rate of flow through them. The holes may in time become completely blocked. The water now finds its way down by overflowing from the primary trough at one or two points and short-circuiting the entire distribution system. The result is that the effective cooling area of the tower is greatly decreased. The average cooling tower can lose 25% of its cooling capacity within a few days of cleaning, in the peak growth periods. A substratum of sulphate-reducing bacteria usually thrives under the algal film, causing corrosion to the metal bolts used for holding the wooden members together. (For a brief review of the corrosion-resistant metals used for metal fittings in cooling towers see Castleberry[1]).

Festoons of slimy blue-green algae, with numerous other types of microbe (green algae, diatoms, bacteria etc.) embedded in them, will also be found on the lowermost structural members that are continuously wetted, in cooling towers of the natural or induced draught type where light has access to these members. Algae that die (or become detached, are carried to unilluminated or hot parts of the cooling system and then die) add to the organic content of the water or form the basis of slime deposits.

The ubiquitous concrete cooling tower suffers these infections in the same way as other towers, but in addition the structural concrete undergoes attack by sulphuric acid produced by the sulphur-oxidizing bacteria (see below).

The Heat Exchangers and Pipework

In the heat exchange (all-metal) part of the system, both microbial and straightforward electrochemical corrosion are encountered. The flow characteristics determine which type of corrosion predominates, to some extent: in the author's experience, slime-forming microbes tend to colonize those areas where fluid flow is laminar while only "pure" corrosion is found in turbulent areas.

A slime coating is probably initiated in the normal way by bacterial cells anchoring on to minute irregularities on the metal surface. In addition, at low points in the system without much water movement, sludge (inorganic debris such as calcium carbonate scale or corrosion products, as well as particles of rotten wood resulting from fungal decay in the cooling tower, and dead algal material) accumulates; and this forms a matrix in which slime-producing microbes can grow. The thickening slime layer begins to act as an insulating coating which prevents the conduction of heat from the metal to the water, so that cooling efficiency is reduced. Gradually the flow rate drops, and the bore of the pipe may become completely blocked.

Maintenance engineers often refer to this slime as "algae", though, as explained above, algae would normally die in these situations and decompose along with their mucilaginous sheaths to provide nutriment for the other (chemo-organotrophic) microbes in the slime. Certain algae are able to grow chemo-organotrophically in absence of light, however, and these may therefore contribute to slime formation though it is doubtful whether living algae are ever a significant component of slime. Some of the principal slime-forming bacteria (Kelly[2]) are *Bacillus subtilis*, *B. cereus*, and species of *Flavobacterium*, *Aerobacter* and *Pseudomonas;* while a variety of fungi such as *Aspergillus*, *Alternaria*, *Penicillium*, *Monilia* and *Trichoderma* spp., and yeasts such as *Torula* (= *Candida*), are often found.

Within the slime layer, oxygen-deficient and even oxygen-free conditions easily arise since oxygen is taken up by the respiring organisms near the surface of the slime as rapidly as it can diffuse in. This sets up a micro-habitat suitable for the growth of the highly corrosive sulphate-reducing bacteria (see Chapter 3 for a full description of their activities), as well as inducing a differential aeration corrosion cell.

Another type of microbial growth in these situations that gives rise to differential aeration cells and provides conditions for sulphate-reducing bacteria to grow is that produced by the iron bacteria. The genera most commonly found are the filamentous forms *Sphaerotilus*, *Crenothrix* and *Leptothrix* (now all thought to be different forms of the same organism), and the stalk-forming unicellular bacterium *Gallionella;* all oxidize ferrous to ferric ions to obtain energy for growth reactions, and cause thick deposits of ferric hydroxide to accumulate between the organisms. Once a tuft of iron bacteria has formed on the surface of a pipe etc., pitting corrosion can proceed under it almost independently of the biochemical activity of the bacteria – the organisms and their ferric hydroxide deposit appear to present a mechanical barrier to the ingress of oxygen. Moreover, the different metals employed in cooling systems all seem to suffer from corrosion under deposits of slime or iron bacteria. Copper evidently does not exert any bactericidal or even fungicidal action in this context, and is attacked. Galvanized and cadmium coated steels are less susceptible than unprotected mild steel, but do suffer from attack. Stainless steels are fairly immune, in the author's experience, in oxygen-free situations but vulnerable in aerobic conditions; however they are very rarely employed in cooling systems.

A heavy growth of micro-organisms tends to render the cooling water acidic and hence more aggressive towards metals, as well as having the more specific local effects described above. The use of phosphate makes the situation worse by accelerating growth, particularly of the algae, which appear to utilize polyphosphates as readily as orthophosphate.

Use of Biocides in Recirculating Cooling Systems

As already explained, the only feasible method of controlling microbial growth in cooling water systems is by the use of antimicrobial chemicals or *biocides*. It is not necessary to render the system utterly sterile (indeed it would be impossible), but merely to keep microbial growth down to a level that does not interfere with heat transfer in condensers etc. and efficient water cooling in the tower, or cause the local onset of oxygen-free conditions. The ideal biocide would meet all of the requirements listed below; the ones used today do not, or at least we lack precise information about these properties in most cases.

Desirable properties of a biocide:

1. Cheap to use at an effective concentration.

Fig. 4.1 *Algal growth inside a cooling tower*
The walkway is set over the primary trough, at the top of the tower. Some of the timbers on the left have been wiped free of growth with a cloth. The tower had been scrubbed clean three weeks previously.

Fig. 4.2 *Section of unprotected spun iron pipe at end of corrosion experiment*
The adhering soil has been removed from the pipe, and the nodulation is clearly visible.

2. Safe to handle (not injurious to skin; not spontaneously inflammable or explosive).
3. Non-toxic to higher forms of life at the concentration used.
4. Broad spectrum of activity against microbes, *i.e.*, controls all troublesome species.
5. Non-reactive towards dissolved organic substances and dead organic matter generally.
6. Non-injurious to metals and wood.
7. Non-volatile to avoid losses in cooling tower.
8. Stable in dilute solution over long periods, surviving repeated heating and exposure to light.

Chlorine has undoubtedly been used industrially more than any other biocide, and is effective at extremely low concentrations (1 part per million free chlorine or less). Like other oxidizing agents, though, it reacts with all organic matter; but in systems that can be kept relatively free from such organic detritus it is an economical method of control. Its use is fraught with hazards, however: except in extremely low concentrations it is not only injurious to metals but it delignifies wood, so that fibres become detached from the surface and accumulate elsewhere in the system to cause blockages or become focal points for corrosion. Moreover, chlorine does not properly control the microbial attack of wood by fungi. Its volatility results in considerable losses in the cooling tower, so that either continuous chlorination or intermittent application at a fairly high frequency is necessary. The latter method is more commonly employed: it is of course more economical and adequate control of fouling can usually be maintained in this way.

The programme of chlorination designed to maintain the necessary residual free chlorine in the water varies with the system: no two are alike in this respect and the correct treatment must be found by trial and error. In general, a residual of about 0·5 p.p.m. free chlorine will usually perpetuate satisfactory conditions in the system, though it may be difficult to ensure that this level is being maintained in all parts of the system. The high B.O.D. of most industrial waters means that one often has to add chlorine to 200–300 p.p.m. to achieve a level of 0·5 p.p.m. free unreacted chlorine in solution. Such treatment will lead to an undesirable increase in the total electrolyte content of the water, which increases the corrosion hazard: one has traded a microbial corrosion for one with a chemical basis. In waters of low B.O.D. chlorine is of great value, but such waters are not common in industry. Regular chemical analyses should be made to ensure that the chlorine level is correct, and this could be combined with conductimetric testing to check that the total dissolved matter in the water is not excessive.

Before leaving inorganic substances, mention should be made of chromate, which is frequently employed as a corrosion inhibitor and has biocidal properties.

Recently, the trend has been towards using organic substances as biocides, and several special formulations have been developed. Their higher price is offset by their relative harmlessness to the system, which does away with the need for special dosing equipment and constant monitoring.

Chlorinated phenols have been widely used. The sodium salts of 2,4,5-trichlorophenol and pentachlorophenol are perhaps the most commonly used; a formulation containing both of these is inhibitory to a wide range of troublesome microbes, including fungi. There is a definite hazard in handling the chlorinated phenols, though; and the use of the above-mentioned ones is forbidden in some European countries because of their toxicity. This of course raises a problem in connexion with the disposal of purge or bleed-off water: at best, wild-life, both plant and animal, is at risk.

Quaternary ammonium compounds have been successfully employed, despite their being relatively innocuous to fungi and less effective in presence of protein and high concentrations of ions. Their ability to penetrate thick microbial growths and break up slime renders them useful, and their effect of increasing the permeability of microbial cells enables them to be used in conjunction with other biocides (*e.g.*, copper salts, which are very active against algae) to increase their penetration into cells. Some quaternaries are known to have corrosion-inhibiting characteristics.

The commercial preparation "Panacide" (dichlorophen) is highly effective; it is safe to handle and is harmless to man and animals at concentrations lethal to microbes. Another effective substance, rather hazardous to handle and with high toxicity, is acrolein: it is said to function as a broad-spectrum biocide at 0·2–1·0 p.p.m. A number of other substances, such as methyl-1,2-dibromopropionate and peracetic acid, have been employed occasionally and apparently very successfully.

In the case of all organic formulations, the manufacturer should be asked to give written assurance that his product is non-toxic and will not introduce any hazards to the neighbouring community or wildlife. A reputable manufacturer will also give a full technical advisory service to support his product.

Almost as important as the biocide is the method of applying it to the water in the installation. It is necessary first to determine what concentration is needed to cope with the microflora in a given situation, usually by determining the M.I.C. (minimum inhibitory concentration) of the biocide with respect to the organisms actually present in the system. Published results of work on industrially important micro-organisms are unfortunately rare, and widely scattered in the literature. The effects of about 200 inhibitory substances on the sulphate-reducing bacteria were collated by Saleh, Macpherson and Miller,[3] but there appears to be no comparable compilation of data on other troublesome species. It is very easy to cause resistant strains of microbes to arise by injudicious dosing schemes; care must therefore be taken to avoid ineffective treatment programmes, though the competent supplier will advise the prospective user how to prevent such conditions from arising. The treatment can then be optimized, largely on a trial-and-error basis, to obtain a suitable degree of control (sterility, as already stated, is unnecessary and virtually unattainable) at minimum cost. In general, a shock-dose programme is preferable since maintenance of a constant biocide level is the surest way of breeding resistant strains; and in a well-run régime, purging routines will be co-ordinated with the dosing scheme so as to avoid their coinciding with the killing periods.

Combined Action of the Sulphur Bacteria in Metallic Corrosion

The Sulphur Cycle

Amongst the biochemical cycles occurring in nature the nitrogen cycle and the carbon cycle are familiar to most people, but less is heard of the equally important *sulphur cycle* and its role in corrosion in industry. This is a continuous cycle of biochemical change which can be brought about entirely by the action of a mixed bag of microbes collectively called the sulphur bacteria, or can be diverted through autotrophic (photosynthetic) plants which incorporate the sulphur of sulphate into their protein (see Fig. 4.3). Plants either die or are consumed by animals, and either way the protein is eventually

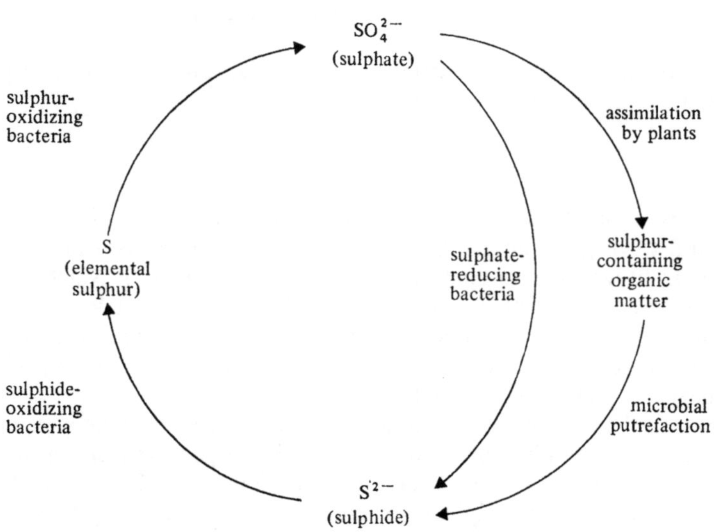

Fig. 4.3 *The sulphur cycle*

degraded by putrefactive microbes to return the sulphur to a simple, reduced form.

The "short-circuiting" action in the cycle, bacterial sulphate reduction, is described in detail in Chapter 3 along with its significance in corrosion. The oxidizing part of the cycle is, however, of equal importance in the present context.

The most common genus of the *sulphur-oxidizing bacteria* is *Thiobacillus*, which is responsible for a significant proportion of industrial corrosion. Sulphur-oxidizing bacteria perform their metabolic functions aerobically (with one exception) on inorganic materials often generated by sulphate-reducing bacteria; they do not require organic matter as a food source (thus they are chemolithotrophs – see Chapter 1) and some species can withstand quite extreme conditions of acidity. These will grow readily at pH 1·0, but will seldom find this necessary since most of the acid which they produce will be rapidly neutralized in attacking metals or concrete.

In aerobic conditions Thiobacilli probably oxidize sulphide, thiosulphate or sulphur according to the following equations:

1. $2H_2S + 2O_2 \rightarrow H_2S_2O_3 + H_2O$
2. $5Na_2S_2O_3 + 4O_2 + H_2O \rightarrow 5Na_2SO_4 + H_2SO_4 + 4S$
3. $4S + 6O_2 + 4H_2O \rightarrow 4H_2SO_4$

Thiobacillus thioparus can also act on polythionates and mercaptans to form sulphate ions. Two very similar species, *T. thio-oxidans* and *T. concretivorus*, are responsible for the formation of free sulphuric acid. Another species, *T. denitrificans*, is a facultative anaerobe: in absence of oxygen it can oxidize sulphur compounds to sulphuric acid by reducing the nitrate ion to gaseous nitrogen.

Two common corrosive situations in which the sulphur cycle is suspected to operate will now be described.

Corrosion in Gas Holders

A good example of the sulphur cycle in action is the case of the infection of coal- and town-gas holders. The water-sealed gas holder is a familiar sight in Britain, though less so than formerly due to the advent of North Sea Gas. The gas holder is, in effect, an inverted bell floating in a tank of water; the tank may contain several million gallons of water which will not be renewed for many years. Overflow from rainfall is conducted away via an overflow pipe near the surface of the holder water.

Towards the end of the last century it was noticed that the water in the holder could contain hydrogen sulphide which was generated *in situ* and then picked up by the gas or caused corrosion on the metal surfaces of the holder; treatment with ferrous sulphate to remove the H_2S as precipitated ferrous sulphide was suggested. Subsequent work, much of it at the then National Chemical Laboratory, Teddington in a unit which has since, rather surprisingly, been disbanded, showed that the H_2S was generated by sulphate-reducing bacteria. These require only a few parts per million of sulphate ions in order to operate to a very high nuisance value. They are, in general, motile and might therefore be expected to be fairly evenly distributed throughout the whole body of water in the holder tank, but, since they are obligate anaerobes, they tend to congregate at the bottom of the tank amongst the sediment, where perhaps reducing conditions can become more intense and thus more favourable for growth (they are very sensitive to the redox potential of their environment).

With the proliferation of bacterial cells an H_2S-rich layer is formed amongst the sludge at the bottom of the tank; much of the H_2S becomes adsorbed on the particulate matter which abounds there. The reaction eventually reaches equilibrium and a period of stasis results. This, in biological terms, amounts to autointoxication of the sulphate-reducing bacteria and is reversible:

$$SO_4^{2-} \rightleftharpoons H_2S$$

This state of equilibrium becomes disturbed when the ambient temperature falls because of a sudden frost. The colder, denser upper layers of water sink to the bottom, displacing the sulphide layer upwards and thus increasing the H_2S concentration at the gas/water interface. The H_2S is then partitioned between the water and the gas and produces what the gas engineer calls "dirty gas" which does not comply with the very strict legal requirements for town gas. Dirty gas is the major effect of this phenomenon but, at the same time, the H_2S also makes its familiar contribution to the corrosion picture of the holder bell and tank – this occurs mainly on the inside since the exterior (periodically exposed) surface is usually painted for aesthetic reasons.

It has long been the practice of gas engineers to seal off the air/water interface at the annulus of the holder with an oil layer, and for many years they were misled into thinking that this would prevent ingress of air and cause oxygen-free (still frequently equated with "corrosion-free") conditions to prevail. In recent years, however, the author and collaborators have found that these oil films appear to have the remarkable property of *inducing* atmospheric oxygen uptake by the water layers which they cover. Whether this oxygen uptake is physical or biological we cannot at present say, but it allows for oxidizing conditions to be present at the top of the holder tank. We have seldom found sulphate-reducing bacteria more than 7 ft from the bottom of the holder tank and there is evidence that the oxidizing portion of the sulphur cycle operates at the top layers regardless of the presence of an oil layer. This ensures a constant recycling of sulphur compounds and makes the uptake of sulphur dioxide from atmospheric pollution less important to the process.

Over the years various methods have been employed to counter this most undesirable contamination of stored gas, most of which were based on the chemistry of hydrogen sulphide; oxidizing, precipitating and "starvation" techniques were used.

All the chemical conversion techniques were merely symptomatic treatments which could not have any long term benefit. Oxidation with chlorine, hypochlorites, hydrogen peroxide and aeration has a purely ephemeral effect, since the resultant sulphates simply set the above equation moving from left to right again. Precipitation with ferrous sulphate, zinc acetate, zinc oxide, copper sulphate and barium chloride has been tried but, with the knowledge that the problem is microbiological, such methods can be seen to be illogical because they either do not reduce the total sulphur content or may even increase it. At the concentrations relevant for bacterial metabolism it is far from certain that any of the "insoluble" products so formed are sufficiently so: zinc sulphide, for example, has an aqueous solubility of 6·9 p.p.m. at 18 °C and even barium sulphate has a solubility of almost 3 p.p.m. at 30 °C.

The starvation technique allowed bacterial growth to continue unchecked until all the SO_4^{2-} ions were converted to S^{2-} and was the first sign that a biological philosophy was being applied. Nevertheless the weaknesses of this approach will be evident to the reader from what has already been said.

The two methods that are rational and effective are the use of the bactericide Panacide or chromic acid. Chromic acid needs careful analytical control and can prove corrosive in its own right if not kept at the correct level; it will also prove expensive if no local supply of waste chromate liquors is available. The bactericide has no such drawbacks.

Some years ago the idea became current that the familiar "rust" stain on the outer surface of the bell of the gas holder was due to accretions of iron bacteria. When we examined this in detail we found that, in most cases, the paint film was quite intact, thus ruling out the possibility of corrosion through the paint film. Microscopic examination failed to reveal any bacterial cells or cell debris, and the deposit was identified as iron oxide. We were forced to the conclusion that the oxide deposit came about by the atmospheric oxidation of iron sulphide films which were picked up by the telescopic "lifts" of the holder when immersed in the tank water. This indicates that corrosion products resulting from attack on the internal surfaces of the holder tank and bell were being transferred to produce an apparent corrosion elsewhere. This theory was confirmed to some extent by the fact that oxide deposition was never found on those lifts which seldom became immersed in the water.

This type of problem is by no means the sole prerogative of the gas industry, for it will be found wherever the appropriate conditions prevail; its inclusion shows the widespread nature of microbially promoted corrosion in an industrialized society.

Pipeline Corrosion

In this section on the combined action of sulphur bacteria it may be of interest to summarize the results of an experimental burial of pipe sections by Purkiss and Cameron[4] in a situation which, though about 1 ft below the mean water table, was subject to fluctuating conditions not always favourable to the activities of the most active enemy of underground pipes – the sulphate-reducing bacteria. These bacteria are commonly the principal cause of graphitic corrosion of spun iron gas mains but it is not uncommon for the whole sulphur cycle to be operating where conditions are favourable – or unfavourable, according to one's viewpoint. The heterogeneous nature of metallic objects causes areas of differing e.m.f. when an aqueous solution of an electrolyte is present, as explained in Chapter 3. The electrochemical corrosion which ensues would normally cease under oxygen-free conditions, when polarization by atomic hydrogen occurs at the cathodic sites. Sulphate-reducing bacteria ensure that such polarization does not happen and corrosion proceeds unhindered, resulting in pitting of steel pipes and "graphitization" of iron pipes. The soil in the vicinity of such a corroding pipe is often blackened and contains a high concentration of sulphide ions.

Where the texture of the soil is sufficiently open to permit free ingress of oxygen, at least under dry conditions, the complete sulphur cycle may operate. In the trial referred to above the presence of a strain of *T. denitrificans* and of *T. thio-oxidans* was demonstrated. Atmospheric oxygen and soluble soil nitrates provided the oxygen source and mineral sulphates the substrate for the full sulphur cycle. This introduced the possibility that, instead of the cycle operating on the basis of a co-existent series of growth reactions, there may be a *sequence* of phases of a broadly cyclic nature taking place. In such a case one portion of the bacterial cycle would be dominant at any given time, passing to another phase when the conditions brought about by the bacterial activity caused the temporary cessation of growth of that type of organism.

Details of the experimental pipe burial by Purkiss and Cameron, and the results obtained, are as follows. A 100-ft trench, divided into 17 compartments by cross walls of compacted clay, was prepared, and a short control trench divided into 2 compartments was made parallel to it and 22 ft away. Sections, 2 ft long, of spun iron gas pipe were laid in the compartments with the following treatments: pipes for which the backfill had been soaked with a 5% aqueous solution of Panacide were coded B; pipes wrapped in plain, non-impregnated hessian, H; pipes wrapped in 2% Panacide-impregnated hessian, I; pipes painted with 40% Panacide solution and allowed to air-dry before laying, P; controls, C. Certain combinations of the above treatments were also employed; all treatments were carried out in duplicate. The backfill in each case was inoculated with a suspension of sulphate-reducing bacteria obtained by cultivating organisms originally isolated from the site. This procedure was to minimize the initial inactivity of sulphate-reducing bacteria due to the permeable nature of a fresh backfill.

Pipe sections were removed and examined over the course of 4 years;

bacteriological tests were carried out on soil samples taken from the immediate vicinity of the pipe sections.

Table 4.1 shows the general conditions in the burial site (which was a heavy grey clay usually in a very boggy condition, and already known to be aggressive), determined from a series of soil samples taken along the line of the proposed trench. Tables 4.2–4.5 show the results of soil tests and visual inspections of the corrosion of exhumed pipes after various periods of burial.

Table 4.1. *Soil conditions at the pipe burial site*

Sample	From hole No.	Depth in ft	Moisture %	pH	Total sulphur p.p.m.	Water soluble sulphate p.p.m.
A	1	3	27·9	8·40	290	63
B	1	5	27·1	8·43	560	115
C	2	3	22·6	8·25	210	20
D	2	5	22·1	8·35	90	28
E	3	3	22·9	8·40	—	—
F	3	5	22·9	7·75	—	negligible

Sulphate-reducing bacteria and *Thiobacillus denitrificans* were identified in all these samples.

Table 4.2. *Soil conditions and corrosion of pipes after 5 months' burial*

Samples of soil from the immediate vicinity of the pipes were put into screw-capped bottles and tested in the laboratory for sulphide and for sulphate-reducing bacteria.

Pipe Code No.	H_2S	Sulphate-reducing bacteria	Corrosion
C1	positive	positive	Slight, mainly at the hack-sawn edges of the pipe
C6	positive	positive	One or two minor graphitic pits
I1	absent	absent	nil
P1	absent	absent	nil
B1	positive	positive	nil
IB1	absent	absent	nil

Table 4.3. *Soil conditions and corrosion of pipes after 15 months' burial*

Pipe Code No.	H_2S	Sulphate-reducing bacteria	Corrosion
PB1	positive	positive	nil
C2	positive	positive	marked
H2	positive	positive	slight at edges
C6	positive	positive	definite

Table 4.4. *Soil conditions and corrosion of pipes after 2 years 5 months' burial*

Pipe Code No.	H_2S	Sulphate-reducing bacteria	Corrosion
C5	heavy	positive	nodules present
H1	negative	positive	slight. Hessian completely decomposed
C4	heavy	positive	nodules present
PB2	positive	positive	slight at edges
I2	nil	nil	no corrosion
P2	positive	positive	very slight: one nodule present

Table 4.5. *Soil conditions and corrosion of pipes after 3 years 10 months' burial*

Pipe Code No.	H_2S	Sulphate-reducing bacteria	Corrosion
IB2	negative	negative	no corrosion
B2	positive	positive	three nodules
C3	heavy	positive	severe: many nodules

By the second batch of exhumations, it was beginning to become clear that the untreated controls (coded C in the Tables) were corroding, and that the drench treatment of the backfill (B) was no longer effective in preventing this.

In the third batch, formation of nodules (first noticed in rudimentary form on pipes of the second batch) had advanced significantly. The untreated hessian wrap was in an advanced state of decomposition owing to the activity of cellulose-splitting organisms; the Panacide-impregnated hessian wrap, and the pipe inside it, were in perfect condition.

The fourth and final batch of exhumed pipes confirmed the previous observations: the provision of a reservoir of Panacide in close proximity to the pipe constituted the best method tested for controlling bacterial corrosion of pipes. The control was very heavily nodulated by this time; Fig. 4.2 shows that a considerable degree of nodulation had developed on the pipe originally covered with untreated hessian but which had by then completely disappeared.

These corrosion nodules appeared as small clay-coloured lumps dotted in random manner over the surface of the pipe. On dissection they were found to be composed of a series of strata as follows:

1. The outermost stratum consisted of an adhering layer of the blue-grey clay.
2. Underlying this was a layer of red iron oxide, usually 1–2 mm in depth.
3. Below this was a layer of black graphite with the typical greasy feel normally found with graphite. This stratum continued down to the original surface of the pipe and considerably below it.
4. Below the nodule was therefore a concave indentation into the metal of the pipe. The graphite could be quite easily lifted with the finger nail revealing these indentations or pits, some of which were 3 mm deep and had a diameter of 1 cm.

Sulphate-reducing bacteria were found in each of these layers of the dissected nodules.

Corrosion in Central Heating Systems

Corrosion in these systems is often a straightforward electrochemical problem, but occasionally – especially in Holland – cases have come to light in which bacterial action is clearly responsible. In an affected installation one often finds an appreciable sulphide content in the water, with deposits of iron sulphide at the points of leakage. The water will also contain some suspended iron oxide and the B.O.D. will be quite high compared with main water. Investigation shows that there is a seasonal variation in the corrosion picture. In a newly-commissioned installation iron bacteria will often be dominant, but in summer when the system is shut down the oxygen in it is slowly consumed by various processes and conditions become favourable for the growth of the sulphate-reducing bacteria. This results in the formation of local deposits of ferrous sulphide, and eventual perforation at these sites which can occur in as little as two years. Organisms recovered from these systems include *Desulfovibrio*, *Leptothrix* and *Thiobacillus* genera (no assignment to species was attempted), indicating that the microbial action could be complex and could involve the sulphur cycle.

Bacterial Damage in the Wire-drawing Process

Wire-drawing is of great importance, particularly in cable manufacture. In reducing the diameter of copper rod to form wire, the rod is forced through a tungsten carbide die which is lubricated and cooled by an aqueous solution of the soaps of various fatty acids. It is noticeable from time to time that the life of a die is considerably below the normal. On investigating such a case it was found that die life could be related to the number of bacteria present in the coolant/lubricant fluid and this provided the major clue. The bacteria cause hydrolysis of the sodium soaps, liberating free fatty acids which can react with copper ions from the surface of the wire passing through the die. As long as these blue copper soaps remain stable they cause no trouble, but, as the concentration of bacteria increases, there is an oxidative attack upon the fatty acid radicals with the precipitation of oxides of copper. Thus the coolant, initially a milky-white, gradually becomes bright blue as the copper soaps form; following this it turns grey, and eventually black, as the concentration of copper oxide increases. It was shown that this oxide content of the coolant was responsible for the additional abrasive action on the die and the consequent shortening of its life. Copper oxide added deliberately to such a system was found to decrease die life in the same way. Control of the growth of bacteria by means of a biocide proved to be the answer to the problem.

Microbial Corrosion in Presence of Oils and Emulsions

This topic is dealt with in detail in Chapter 5, but brief mention may be made here of two examples of this type of corrosion.

The first recorded observation of microbial growth on a liquid hydrocarbon appears to have been made in 1895 by Miyoshi,[5] but only recently has its metallurgical significance been realized. The mechanism of utilization of this unusual kind of substrate is described in Chapter 1, and will not be discussed here; suffice it to say that the presence of water is necessary in the oil.

Compressed air installations usually contain appreciable amounts of oil, to the extent that water trapping devices may have to be fitted; droplets of lubricating oil from the compressor unit find their way into the traps. Needless to say, the conditions are predominantly aerobic though not exclusively so.

The water in these systems will contain fair numbers of iron bacteria sheathed with ferric hydroxide; various pseudomonads such as *Pseudomonas oleovorans* and *P. fluorescens* may be found. In common with many other microbes, certain pseudomonads are able to degrade saturated hydrocarbons such as lubricating oils, producing acidic end-products which initiate corrosion. This is a type of corrosion in which the sulphate-reducing bacteria do not appear to play a major role.

We have recently shown (Littler and Purkiss[6]) that suppression of microbial growth can be achieved in a cutting emulsion used in a screw-cutting plant by maintaining a concentration of 400 p.p.m. of Panacide C.A. in the emulsion. It was calculated that a saving of £8000 per annum would result from this treatment because of the decreased reject rate due to poor cutting and acid-staining of work, and decreased frequency of emulsion change. In addition, bad smells were eliminated, and the problem of disposing of spent emulsion became much less pressing.

The Role of the Biotechnologist

The Biotechnologist's Role in Treatment

The conventional bacteriologist is accustomed to working in situations where the problem, though critical, is of a relatively uncomplicated nature. He seeks to identify a single causal microbial species as the agent of the trouble: for example, he confirms a diagnosis of tuberculosis when he has found significant numbers of *Mycobacterium tuberculosis* in the sputum. The industrial microbiological problem however is *hardly ever* uncomplicated, though fortunately it is less critical. It generally involves a complex microflora of many genera and species, affecting each other chemically and physically. Some of the microbes will be found to be acting in an atypical role or synergistically – or merely in sequence with others. It is seldom of much value to spend time in attempting to identify accurately the constituent members of the microfloral population: it is very much more to the point to establish means for the cure of the infection. This might be done by establishing the minimum inhibitory concentration (M.I.C.) of a group of acceptable biocides towards the responsible organisms and then to apply the principles outlined earlier in this chapter under the heading *Use of Biocides in Recirculating Cooling Systems*.

Another form of rational action in some situations would be to investigate the source of infection and take the necessary steps to eliminate it or seal it off.

A curious feature of industrial infections is the manner in which a microorganism – even the dominant one – in the population can disappear, giving place to another, without there being any significant alteration in the gross appearance of the corrosion. The problem remains the same although the major causative organisms may have changed.

Another curious aspect is exhibited in the wire-drawing problem referred to above. A relatively high local temperature will occur at the site of deformation of the metal. We might therefore expect to find a predominance of thermophilic microbes, but in practice we do not. Similarly, the presence of high concentrations of fatty acids would seem to dispose the system to infection by heterotrophic (chemo-organotrophic) organisms, yet the main trouble seems to be caused by autotrophs (chemo-lithotrophs). Again, copper salts are known to be toxic to microbes but all the organisms in this system are patently copper-tolerant – or rather, only copper-tolerant mutants have been permitted by circumstance to survive. Add to this the fact that microbes often tend to be highly specific in respect of the chemical make-up of their environment and that the commercial fatty acids used in the wire-drawing operation are of extremely variable composition and one has some conception of the complexity of the mechanism of this phenomenon.

The Biotechnologist's Role in Design

We have seen how the corrosive activities of microbes can be curbed to a greater or lesser extent by biocide treatment when mere prevention of access or mechanical removal of microbes fails. Some of the problems could be avoided if the design engineers had the necessary information to enable them to foresee the possible biological consequences of some features of the installation. One frequently encounters industrial plant where a perfect environment has been provided for bacterial growth by some unnecessary feature of the design.

A case in point was a large company's cooling system involving a 50,000 gallon storage pool. Cooling was accomplished by passing the warm water into the pool *via* an ornamental fountain. As a final artistic embellishment it was decided to illuminate the pool and fountains at night by immersing fluorescent lights in the water; for a while the effect was attractive. Very soon there appeared a heavy algal growth which soon found its way into the heat exchangers and gave rise to slime formation; the pool itself soon assumed all the beauty of a village duckpond without the good offices of the ducks to scavenge the organic growths. Eventually the whole system had to be replaced. This design was the natural result of lack of appropriate knowledge, and illustrates how the biotechnologist can play an important part at the design stage.

If an installation is built which will contain or make use of water, it must be realized that microbial infection will inevitably take place sooner or later. Even if the water is merely used as a solvent or vehicle for some organic agent, the same risk of infection will always be present. It is essential to consider at the design stage what the microbial implications of a design feature will be and, where possible, to take the necessary evasive action. Just what that action should be is difficult to generalize upon, but the contributors to this work can hope to have sign-posted the lines on which thinking should be directed.

* * *

Fig. 4.2, and Tables 4.1 to 4.5, are reproduced by kind permission of the Editor, *Gas Journal*.

References

[1]CASTLEBERRY, J. R. '*Trends in cooling tower construction*'. Materials Protection, **8** (3), 67–70 (1969).

[2]KELLY, B. J. '*New chemical formulation will control micro-organisms in hydrocarbon contaminated cooling water systems.*' Materials Protection, **4** (7), 62–67 (1965).

[3]SALEH, A. M., MACPHERSON, R. AND MILLER, J. D. A. '*The effect of inhibitors on sulphate-reducing bacteria: a compilation.*' J. Applied Bacteriol., **27**, 281–293 (1964).

[4]PURKISS, B. E. AND CAMERON, R. W. N. '*Pipeline corrosion inhibited by a bactericide*'. Gas Journal Supplement, 3–8 (Sept.–Oct. 1965).

[5]MIYOSHI, M. '*Die Durchbohrung von Membranen durch Pilzfäden*'. Jahrbücher für Wissenschaftliche Botanik, **28**, 269–289 (1895).

[6]LITTLER, E. H. AND PURKISS, B. E. '*The control of biodegradation of cutting fluid*'. Industrial Lubrication and Tribology, **22**, 108–111 (1970).

CHAPTER FIVE: E. C. HILL

Microbial Infections in Relation to Corrosion During Metal Machining and Deformation

Infection of Emulsions

General Considerations

Micro-organisms are involved in corrosion in metal machining and deforming operations in a variety of ways. Their role is often largely indirect and is due then to their destruction of anticorrosive agents in the coolants used. These coolants are often oil emulsions in which the good cooling properties of water are married to the lubricant and anticorrosive properties of oil and additives. Emulsions used for machining operations are chemically stabilized dispersions of oil in water, the stability being achieved by the use of surface-active anionic or non-ionic emulsifying agents. Emulsions used for some other purposes, for example as hydraulic fluids, where flame-proofing rather than cooling is important, are frequently stabilized water-in-oil dispersions.

Some degree of infection is almost always present even in a fresh charge of emulsion, as the coolant quickly acquires an inoculum of organisms from the slime and deposits in the sump and pipes of the plant. Growth and reproduction of the organisms occurs rapidly and within a few days millions of microbes per ml are often present.

Eventually a state of balance tends to be reached in which the population remains steady, for although some reproduction still goes on, and more organisms may arrive from infected workpieces, dirt etc., there is a continuous loss of organisms from the system due to their removal by filters or centrifuges, carry-over on machined parts, death by local heating and dilution by top-up.

In oil emulsions the alkaline pH favours the growth of bacteria, such as species of *Pseudomonas*, *Alcaligenes*, *Proteus*, *Vibrio* and *Achromobacter*; but it is relatively common for the pH to fall to a level where yeast growth is encouraged (Fig. 1.9 shows typical yeasts). Sometimes the growth of yeasts is inadvertently selected for if a narrow spectrum biocide, such as tris-(hydroxymethyl)nitromethane which is effective against bacteria but not yeasts, is employed.

When freshly made up and circulated, an emulsion contains dissolved oxygen which favours aerobic bacteria, but the oxygen demand of these microbes soon causes stagnant parts of the system to become oxygen-deficient, particularly in the presence of metal fines, and anaerobes such as the sulphate-reducing bacteria may flourish locally. Oxygen depletion is favoured by any shut-down period, and since the anaerobes tend to be the smell producers, restarting a machine on a Monday frequently releases an embarrassing odour (the so-called "Monday morning odours"). In a machine tool coolant the in-use temperature of the emulsion is usually just above ambient and ideally suited for microbial growth. In a rolling coolant the bulk temperature may be over 50 °C and this encourages a population of thermophilic microbes, often dominated by species of *Nocardia*, which generally speaking are less destructive than those found at slightly lower temperatures (the mesophiles). If the temperature is allowed to drop overnight or at week-ends, the activity of the thermophiles decreases and that of the mesophiles increases.

Thus, in any one system a wide variety of microbes may exist, each most active at a certain temperature, pH and oxygen tension, and each with a different spectrum of biochemical activity.

Growth Requirements

Like all living things, micro-organisms have a dietary requirement for much carbon and lesser amounts of nitrogen and phosphorus, together with traces of many elements listed in Chapter 1. These requirements they satisfy from

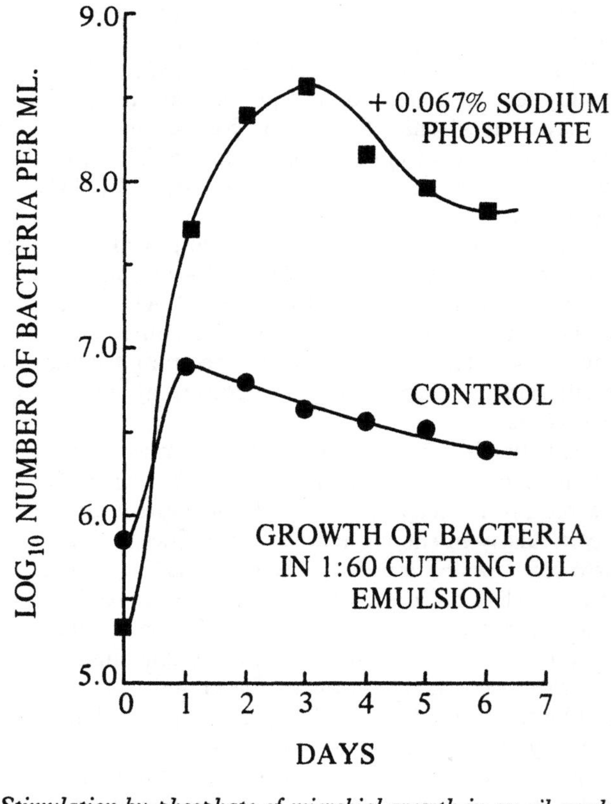

Fig. 5.1 *Stimulation by phosphate of microbial growth in an oil emulsion*

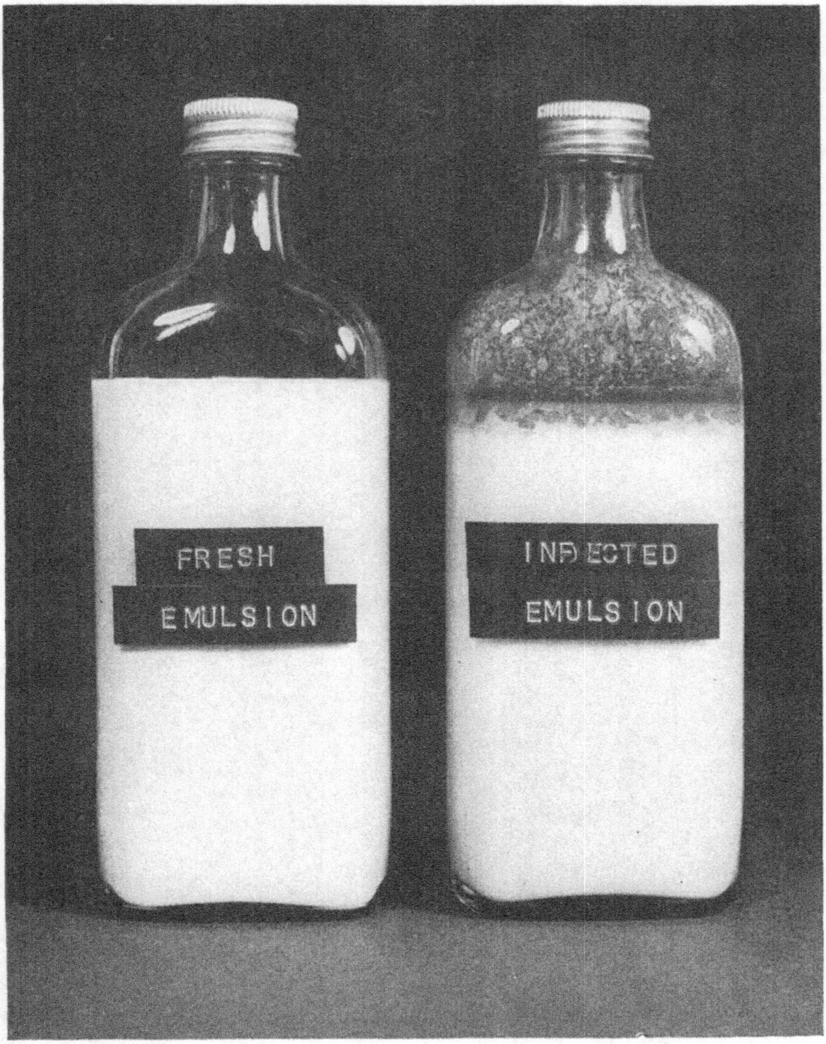

Fig. 5.2 *Fresh and infected aluminium rolling emulsion*

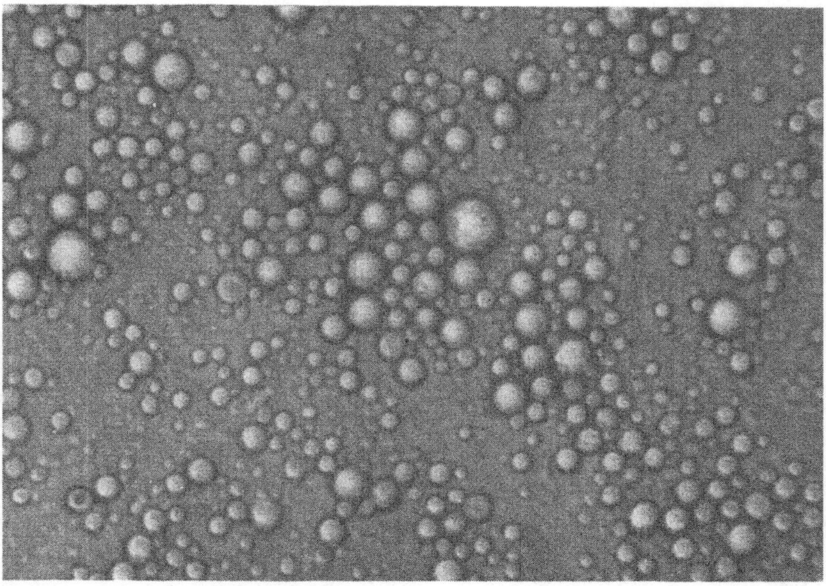

Fig. 5.3 *Photomicrograph of a fairly coarse emulsion*

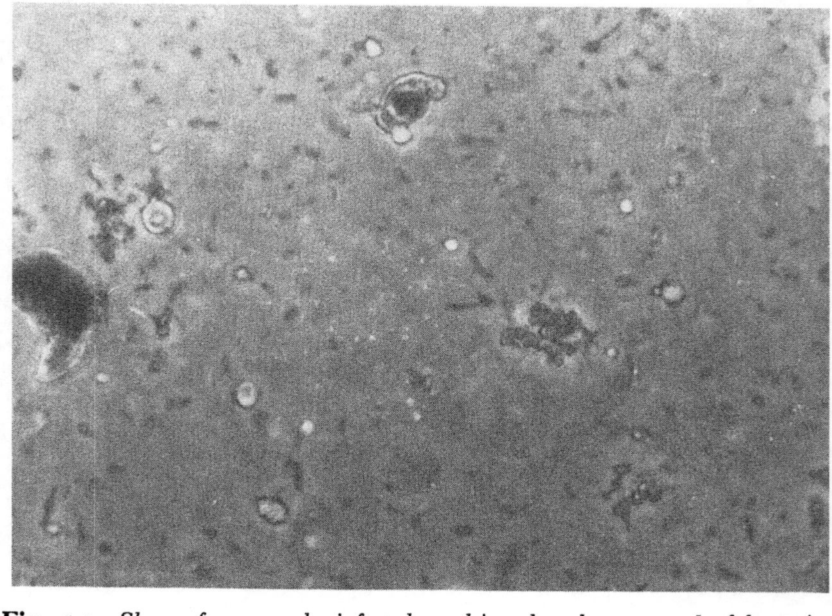

Fig. 5.4 *Slurry from grossly infected emulsion, largely composed of bacteria*

the ingredients of the emulsion formulation, plus extraneous matter such as cigarette-ends, sandwiches, sputum, urine and rags, which find their way into most systems. The microbes live in the aqueous phase and feed on the materials dissolving into it. They do not survive well in the oil phase and hence water-in-oil emulsions tend to suffer less from microbial problems, as the organisms have difficulty in spreading through the continuous oil phase into the dispersed water droplets in which they can grow and reproduce. These water droplets may also contain inhibitory concentrations of chemicals.

Although there is ample carbon for microbial growth, phosphorus and nitrogen, commonly present as constituents of extra-pressure additives and anticorrosive agents, may be limiting, particularly after a large population of organisms has grown up. Fig. 5.1 illustrates the enhancement of bacterial growth when phosphate is added to an emulsion in the laboratory. Whilst it is axiomatic that growth of microbes is accompanied by chemical changes in the environment, a population once it has developed will continue to catalyse chemical changes even when prevented from reproducing by a limiting nutrient, and these chemical changes may in fact be different from those promoted during active growth.

Chemical and Physical Changes Produced by Microbial Growth

Probably the most obvious symptom of chemical change is the progressive separation of free oil from an emulsion due to the degradation of emulsifiers. Fig. 5.2 illustrates this. A population of a few million bacteria per ml will usually bring about this change, but naturally the time element is important and a few bacteria feeding for a long time will achieve as much as many bacteria feeding for a short time. If the process is allowed to continue it is possible to end up with a slurry containing several per cent. of bacteria by weight and having no resemblance to the original oil formulation (Figs. 5.3 and 5.4). Long before free oil separates it is possible to note stability changes of the emulsion by measuring its capacitance, and to observe microscopically a progressive increase in the size of the dispersed oil droplets. Fig. 5.5 shows the recorder traces obtained from capacitance measurements of a fresh and an infected emulsion using a Wayne Kerr Autobalance Universal Bridge. The emulsions are first mixed in a standard way and the capacitance is then measured continuously. The migration of oil droplets as "creaming" occurs affects capacitance. In a perfectly stable emulsion the tendency to "cream" will be much reduced or non-existent.

The term *oil emulsion* covers a wide variety of chemical formulations used for different purposes at different concentrations and temperatures, and it must be appreciated that each may have a different capability of supporting growth and a different tolerance to growth. Acceptable degradation for one purpose may be intolerable for another.

The progressive increase in corrosiveness of an emulsion during infection

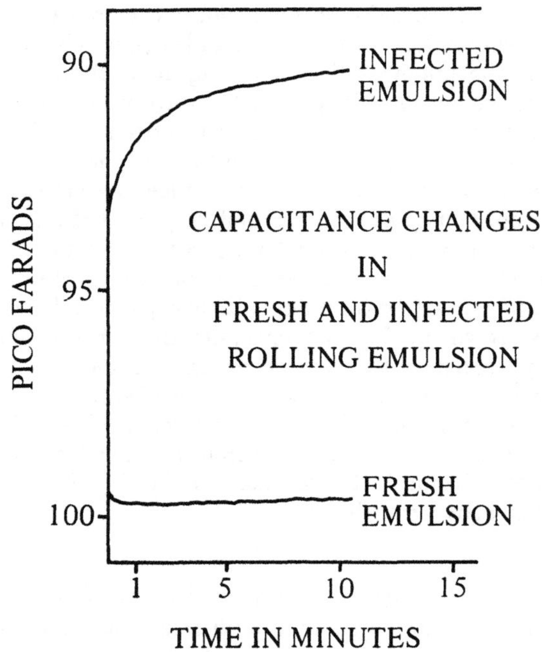

Fig. 5.5 *Capacitance measurements on fresh and infected rolling emulsion over a ten minute period (recorder traces)*

can readily be demonstrated in the laboratory.[1] In brief, the test consists of pipetting 2 ml of the emulsion on to 2 g of standard iron chips (*e.g.*, milled cast-iron Beeny corrosion test plates from Alfred Herbert Ltd., Coventry) on a filter paper (Fig. 5.6) and after two hours' contact time noting the extent of rust staining on the filter paper. If an emulsion normally used at say 3% strength is also tested at greater dilution, there will be an obvious increase of corrosion with increasing dilution. This is illustrated in Line A of Fig. 5.7. If the 3% emulsion is then infected and re-examined at intervals by making dilutions and testing each with the 2-hour corrosion test (Fig. 5.7, Line B), it will be clear that the corrosiveness of the emulsion has increased. This is largely due to the destruction of the anticorrosive additives by the infecting organisms. If the test chips are actually made up from the metal being machined in the works, the results are more directly pertinent. A corrosion test, without dilution, of an emulsion in works use can be compared with the result for fresh emulsion (*i.e.*, Fig. 5.7, Line A) and an indication of deterioration obtained. The test is also useful for testing possible biocides to ensure that they themselves do not affect the corrosiveness of the emulsion, and it is

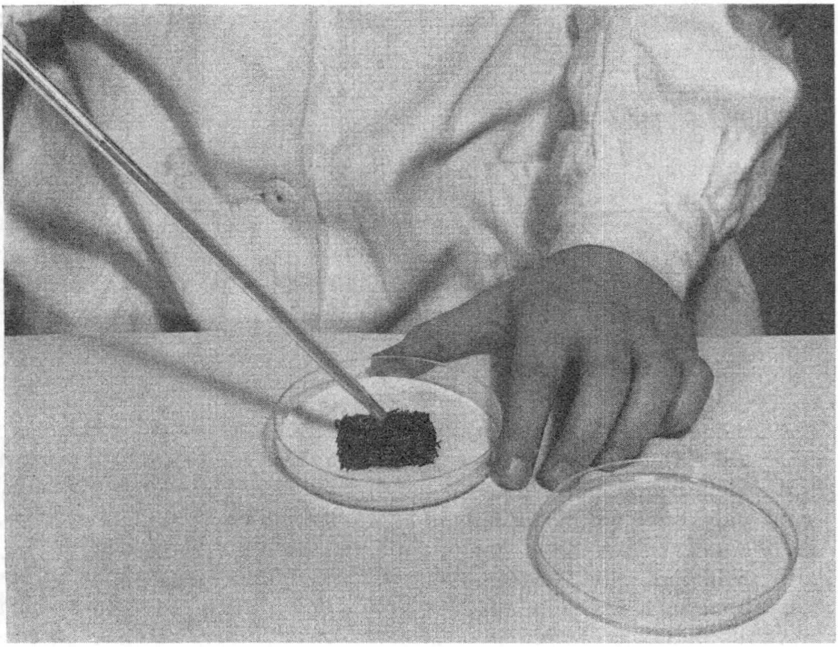

Fig. 5.6 *Pipetting test emulsion on to iron chips for the two-hour corrosion test*

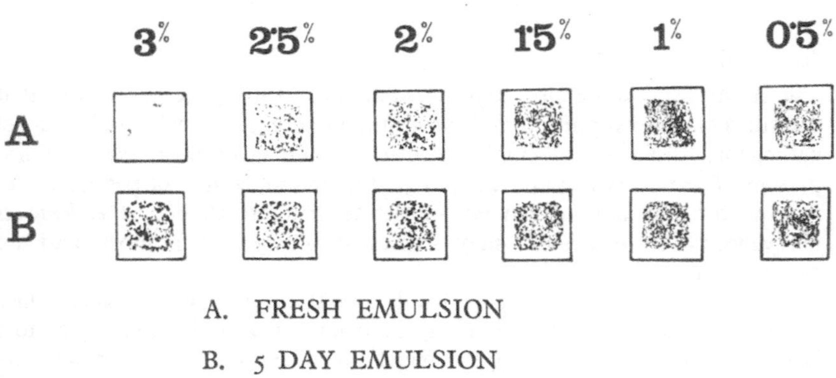

Fig. 5.7 *Actual results (squared-off filter papers) of two-hour corrosion tests on dilutions of fresh and infected cutting emulsion*

logical in addition to test the influence of biocides on capacitance, as most biocides have some effect on emulsion stability.

Synthetic Coolants

Over the last few years there has been a tendency to replace some oil emulsions with synthetic coolants, often largely based on sodium nitrite; but these too suffer massive infections, often resulting in ammonia liberation, and almost always a deterioration in anticorrosive properties. The infected synthetic coolant may become extremely alkaline, with a pH of over 10·0, and of course the ammonia itself may be corrosive to some metals.

Economics of Infection – Machine Tools

The economics of infection of machine tool coolants are probably much more concerned with corrosion of the work-pieces and the tools than with the cost of replacing the coolant. Down-time during a coolant change may be a very significant cost factor, particularly in highly utilized multi-machine systems, and the interruption in the flow of parts may have cost repercussions elsewhere. Although from chemical and physical considerations it would appear that tool-wear would be adversely affected, this is difficult to prove under practical conditions. Infection of grinding coolants may however be correlated with wheel blinding and loss of productive time, and also diamond costs when wheel dressing.

When we add in the additional costs of microbial slime removal by filters and centrifuges and possible time loss or poor production rates due to "Monday morning odours", the total bill resulting from infection of machine tool coolants may reach very substantial proportions.

Industrial Health

The health aspects are also important but it can usually be guaranteed that measures taken to suppress emulsion-degrading organisms will amply control pathogenic (disease-producing) organisms of human origin if they are also present. The majority of emulsion-degrading microbes are not pathogenic to man within our normal concept of the term although one, *Pseudomonas pyocyanea* (= *P. aeruginosa*), may be present in very large numbers and can cause blue pus.

However, as well as having his hands almost continuously immersed in a concentrated suspension of microbes, the machinist will also be exposed to a heavily infected mist (particularly in a grinding operation) which is potentially infectious when inhaled or possibly capable of causing lung sensitization, and it may also be a hazard to the eyes. It seems unlikely that the emulsion-degrading organisms are a primary cause of dermatitis but they may be a contributory factor after the skin has been sensitized or mechanically abraded.[2] Pathogenic organisms originating from the machinists may find their way into, and be disseminated by, the coolant and thus infect other operatives, but generally speaking the emulsion-degrading organisms are actively antagonistic to them and limit their survival.

It is worth pointing out a fallacy in much of the testing of oils for irritant or carcinogenic properties: fresh formulations are tested whilst in practice the formulation is often rapidly changed chemically by infecting organisms. Logically, therefore, infected oils should be tested also. It is unsafe to read too much significance into health aspects in the light of the limitations of our existing factual knowledge. Perhaps more detailed records of a man's work in relation to his health will eventually confirm or refute the above comments.

Economics of Infection – Rolling Mills

The economics of infection of rolling mill coolants are sometimes staggering. Progressive failure of the emulsion can be expected to result in roll-slip and eventually "refusals"; and, in aluminium rolling in particular, infection is associated with "pick-up" which adversely affects the finish on the strip.[3] In steel rolling the presence of iron fines encourages anaerobic bacteria, and these may generate hydrogen sulphide prolifically. This may itself cause a sulphide corrosion stain on the strip and under favourable conditions sulphide-producing organisms can continue to flourish and generate sulphide when deposited in the oil. The substrate for their activity may be sulphate carry-over from the pickle line or sulphur-containing constituents in the oil. Hydrogen sulphide dissolves in hydrocarbons, sometimes with the formation of corrosive organic sulphides, and corrosion can continue at a distance from the site of the original infection. This phenomenon has been documented for jet fuel[4] and there is good reason to believe that it occurs elsewhere.

Whilst the same degradation of anticorrosive ingredients must occur in rolling coolants as in machine tool coolants, and will presumably lessen the protection afforded the plant, the effect on the strip is not known. A microbial phenomenon which does affect the strip has however been demonstrated recently.[5] A heavily infected rolling coolant releases on to the strip flocs of organisms which may adhere there. Subsequent annealing processes are equivalent to a microbial crematorium and the flocs are converted into granules of ash rich in potassium and phosphorus and also sodium, magnesium, calcium and other metals, often as chlorides. The ash granule then becomes the focus of corrosion, giving rise in steel production to "speckled rust". As well as deposits from the incinerated microbes, the chemical changes in the formulation may give rise to non-volatile products which will also contaminate the strip. As will be seen later, this corrosion phenomenon following annealing makes the use of biocides difficult, as biocide treatment of an infected emulsion results in the release of dead microbial slimes on to the strip and the first result will be to make corrosion worse. Prevention, rather than cure, is thus far more important in rolling mills than with machine tools. For example in a rolling mill three fluid systems are involved in the rolling process, namely the hydraulic fluid and the roll-neck bearing lubricant as well as the rolling coolant. All of these may become infected and leakages are normally such that cross-infection can readily occur. All three systems may therefore have to be treated to attain the ideal of prevention of infection of coolant rather than cure.

Infection of "Straight" Oils

Recirculated roll-neck bearing lubricants are particularly prone to infection since they become contaminated with coolant and are maintained at excellent temperatures for microbial growth. Fig. 5.8 illustrates the effect of adding a biocide to an infected Morgan bearing oil: the slimy, turbid appearance of the oil changes, suspended water separates out, and the oil viscosity returns to normal. This ability of micro-organisms to stabilize a water-in-oil dispersion is characteristic, and probably contributes to the corrosiveness of the oil as do the chemical changes which the organisms bring about.

The hydraulic fluid may be an emulsion or a straight oil and in either case may be infected.[6] The pressures involved have little effect on the organisms, and in some plant designs may even assist growth by forcing oxygen into solution. An infected straight hydraulic oil is illustrated in Fig. 5.9.

Infections of straight oils and fuel oils are assuming increasing importance, not least those occurring in jet fuel (Fig. 5.10) where the consequences may be disastrous. Whilst developing acidity is easy to demonstrate in the free water phase in which the organisms live, little is known as yet about the chemical changes in the lubricant or fuel. The tendency of the aqueous phase to have a neutral or acid pH means that fungi and yeasts can flourish as well as bacteria. The species of bacteria are similar to those occurring in emulsions, *Pseudomonas* spp. being particularly common. *Cladosporium resinae* is a common fungal species in fuels, and species of *Penicillium* and *Aspergillus* are also abundant in both fuels and other oils. *Candida* and *Rhodotorula* are common genera of yeasts. Even a trace of water is sufficient to initiate growth, and as water is itself a product of microbial growth on oil more water becomes available to sustain the infection. In some mills the bearing lubricant may be dispersed as an oil mist and not recirculated; the rolling coolant also may be "straight through" and only used once. In these cases biological interventions are naturally minimal.

Infection of Coating Oils

One further microbial involvement in corrosion recently noted is the ability

Fig. 5.8 *Fresh, infected and biocide-treated Morgan bearing lubricant. Note resolution of treated oil into discrete oil and water phases*

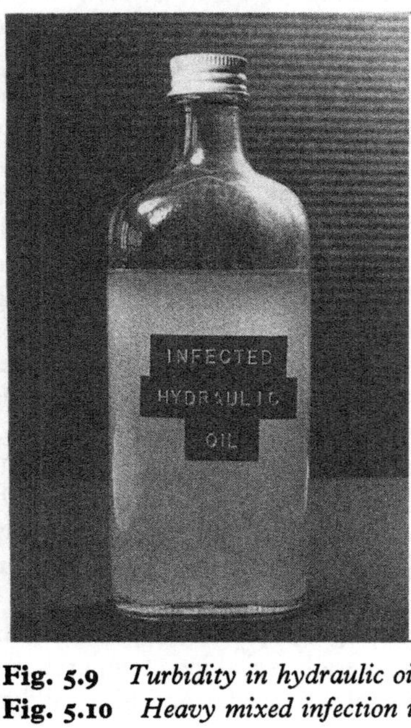

Fig. 5.9 *Turbidity in hydraulic oil caused by microbial infection*
Fig. 5.10 *Heavy mixed infection in jet fuel/water condensate*

of certain organisms to grow in coating oils. These oil formulations usually consist of mineral oil, lanolin and additives and are used to protect machined products or rolled steel from corrosion during handling and storage. The infection may be present in the coating oil system (which may actually smell rancid) or may develop in the coating oil after the strip or machined part has been protected. Fungal infections of coating oils are not uncommon and the metal below the growth usually shows signs of corrosion. The most severe infection yet investigated has been a massive infection of a steel strip coating oil system by yeasts, fungi and bacteria.

Preventive Measures

Good Housekeeping

An appreciation of the biological factors involved in infection leads to fairly obvious precautionary measures. Perhaps the most obvious and important is the complete draining and preferably cleaning and sterilizing of a system at oil or emulsion changes. If we consider the exponential growth of a bacterium with a generation time of 30 minutes, then in say three hours, $1 \rightarrow 2 \rightarrow 4 \rightarrow 8 \rightarrow 16 \rightarrow 32 \rightarrow 64$ organisms, an increment of 63 organisms. Starting with 1000 instead of 1, the increment in the same time would be 63 000 organisms and the chemical change in the growth substrate would be related to the growth of 63 000 organisms instead of 63. Thus it is of prime importance to start a fresh charge of oil or coolant with the lowest practicable infecting inoculum. The worst rolling mill infection in the writer's experience occurred where only half of the coolant was discharged at a time and was replaced by fresh emulsion.

Oils as supplied are usually free from infection but occasionally there are exceptions. Emulsifiable oils are sold containing all required additives as well as emulsifiers and the customer need only mix with water to obtain a usable emulsion. Naturally it is important to use a good quality water in making emulsions. Tap water contains an almost insignificant number of organisms and generally speaking the softer it is chemically the better. Softness becomes particularly important when coolants are kept in use for long periods, as evaporation concentrates the minerals present. It should be borne in mind that de-ionized water may contain organic matter and often a substantial number of bacteria. As a laboratory gimmick emulsions can be made with sterile water and emulsifiable oil in a sterile container and will remain virtually unchanged for years. Unfortunately such measures are impracticable in industry.

Systems should be designed simply, so that stagnant pipe runs cannot occur and extraneous matter can be excluded.[7] Generally speaking, temperatures should be as high as possible in rolling mill coolants, all day and night and at week-ends. Anaerobic conditions lead to offensive odours and undesirable growth products, and continuous aeration is preferable. Cooling towers are excellent fermenters and should be replaced by heat exchangers.

Formulation Changes

Whilst formulations should ideally not comprise a complete microbial diet, this may be difficult to achieve in practice. Even small changes in a formulation may have a marked effect on its microbiology, sometimes in an unexpected way. For example, if the infecting microbes attack the base oil of an emulsion preferentially to the emulsifiers, the product as a whole may be better able to withstand infection as the destruction of a little base oil is of less practical consequence than the destruction of emulsifiers or other additives.

The inclusion of biocides in the oil/additive formulations supplied for make-up into emulsions is difficult, as the biocide must then of necessity be oil-soluble and yet is later required to migrate into the aqueous phase when the emulsion is made up if it is to register activity against infecting organisms. Where the same formulation is used at many different emulsion strengths, it also becomes difficult to select the right biocide concentration.

When considering the oil formulation as a diet it is wise to consider contaminating liquors also. For example, a very severe infection in a marine diesel lubricant has been investigated in which the stimulus for rapid microbial growth was the normal leakage of the piston coolant, a solution of sodium nitrite and phosphate, into the sump oil. The nitrogen and phosphorus continuously provided thus ideally completed the diet for the contaminating microbes.

Physical Sterilization

Whilst steam lances may be of some use in cleaning down and sterilizing a system, heat treatment of infected coolants and lubricants is rarely practical or economic. It is not logical to heat treat a part of the flow and then return it to the infected bulk, as sterility is never achieved. Raising the entire bulk temperature would be the only way of sterilization if it were practicable. It is impossible to put a figure to the temperature required as the time factor involved and the type of infection are relevant. Gamma radiation will undoubtedly sterilize,[8] but undesirable chemical changes may occur in the formulation. Neither sterilization by pressure nor preservation by cold can be considered for industrial systems.

Biocides and Biostats

Both biocides (which kill the unwanted microbes: see the note about the terminology of antimicrobial chemicals in Chapter 1) and biostats (which prevent reproduction) may be useful agents.

Sterilizing Oil Emulsion Systems in Practice

For sterilizing plant at emulsion change a biocide is required. It cannot be effective unless surfaces are cleaned either at the same time or by a previous operation. Perhaps the most satisfying result is obtained when cleaning is done first by the use of conventional degreasants and this is followed by biocide treatment. To some extent both processes can be carried out at the same time, for example by using 0·2% Panacide (B.D.H. Ltd.) in the presence of an alkali cleaner.

The time available for sterilization has to be equated to pumping rates, holding time etc., and there may be a problem of disposal of large volumes of liquid. Biocides usually work very much better at elevated temperature and even a few degrees rise will be beneficial.[9] It must be borne in mind that unless the system can finally be rinsed there will usually be residual biocide left to contaminate the next emulsion charge. For this reason the cationic surface-active biocides, which both clean and sterilize, can rarely be used, as most emulsions are anionic and hence incompatible. If carry-over of biocide into the emulsion is inevitable, compatibility of the two is essential.

It is logical to extend antimicrobial measures to afford protection to the new emulsion charge by adding a biostat (or biocide) so that re-infection is prevented. Practical experience shows that in fact complete sterility is not necessary, as a hundredfold or even tenfold decrease from the normal population of microbes means a proportionate reduction in the problem. It has been pointed out previously that as well as controlling the appropriate micro-organisms a biostat or biocide must not affect the corrosiveness or stability of an emulsion. As there is no "cure-all" in this field, a screening programme will have to be initiated in the laboratory. Ideally this is then followed by a works simulation experiment which lasts for the same period as occurs in the works, and in which the emulsion is alternately aerated and stagnant, in contact with metal fines and possibly heated as well. In this experiment the protected emulsion can be continually challenged by adding micro-organisms, and if infection takes place, further biocide additions made. An apparatus used for this purpose is shown in Fig. 5.11.

Finally a carefully controlled works trial is initiated and actual works experience obtained. Conditions of works use are always more severe than any laboratory simulation and modifications in treatment doses may have to be made. Although it might be thought that a very large initial addition of biocide would give protection for a long time, in practice this would usually have some detrimental effect on the emulsion and would anyhow be depleted by adhering to the metal being machined or rolled, and diluted by topping up.

Small doses over a long period are used instead, often with guidance from rapid microbiological tests.

Of over-riding importance in the final choice of biocide must always be the possible toxicity to the operators, both by skin absorption and inhalation; and obviously cost, ease of handling and availability must also be considered.

Nothing lasts for ever and a microbe-free emulsion will eventually fail for some other reason, such as depletion of additives by oxidation or drag-out, or accumulation of metal fines. Hence the target is never an indestructible emulsion, although it is often found that the control of microbial degradation allows appraisal of physical and chemical degradation to be made with greater clarity.

Finally the biocide-treated charge must be disposed of, and it is advisable to ensure that no problem is occasioned by this. High concentrations of biocide arriving in "slugs" at the local sewage works will certainly not be popular. If emulsions are cracked before disposal it may well be that the biocide is removed at this stage.

Sterilizing Straight Oil Systems

In emulsion systems used as coolants the water phase is continuous and, as it contains the infecting microbes, water-soluble biocides are used. In infected straight oil systems the water phase is dispersed and it is therefore necessary to use a biocide with sufficient hydrocarbon solubility to dissolve in the oil phase and sufficient water solubility to migrate into the water droplets. Some of the chlorinated phenolic compounds have both water and oil solubility and can be finally selected after laboratory trials.

Infections in straight oils lead to stable water-in-oil dispersions due to the surface activity of the living microbes (many microbes secrete proteins or peptides with surface-active properties during growth). The promotion of a stable water-in-oil dispersion by microbes infecting straight oils may seem anomalous as under *Chemical and Physical Changes Produced by Microbial Growth* demulsification was described as a symptom of infection. This demulsification, which occurs in formulated emulsions, is due however to the microbial destruction of chemical emulsifiers.

A good biocide can be expected to resolve this water-in-oil dispersion into discrete water and oil phases (see Fig. 5.8). It is obviously important that the biocide itself should have minimal surface activity or the resolution into discrete phases will not occur. The biocide must also not affect the viscosity of the oil and must otherwise have similar characteristics to those detailed for biocides used in oil emulsions. Experience indicates that prevention is again preferable to cure, and systems should be disinfected between oil charges. This can be achieved by dissolving appropriate biocides in a light paraffinic oil, possibly with the addition of an oil-soluble surfactant, and flushing this through the system for about an hour.

Although somewhat beyond the scope of this article, it is worth mentioning that much of the expertise developed for treating jet aircraft and ground installations is applicable. If a system is known to have a permanent water bottom, it may be sufficient to maintain an inhibitory concentration of a water-soluble biocide in this phase. A convenient method of application is to package the biocide in a polyvinyl alcohol bag. This is insoluble in hydrocarbons but dissolves in water, releasing the biocide. Chromates, often used as corrosion inhibitors, have some antimicrobial properties and can be appropriately blended to produce the maximum effect in a water bottom.

Resistant Organisms

It is always possible that biocide-resistant strains will develop in a system or that a narrow-spectrum biocide will favour the development of a less common organism. It is well, therefore, to have a plan in mind to allow for this, such as routinely changing from biocide A to biocide B and back again or formulating several biocides together. The development of resistant strains is usually associated with sub-optimal levels of treatment, and it is therefore of prime importance to keep the concentration of biocide up. There is no general rule for in-use concentration, as one biocide may be effective at 25 p.p.m. and another at 1000 p.p.m. in one situation and the relative efficiencies reversed in another situation. Even in the same situation the in-use concentration will be influenced by the level of infection: once a gross infection has occurred it may be extremely difficult to regain control.

Methods of Application

There are several approaches to control by biocides. Perhaps the easiest is by programmed shock treatment which has the advantage of discouraging the growth of resistant strains. This method is criticized, however, on the grounds that it permits damaging growth between shock treatments. Control, guided by rapid-test results, has the advantage of suppressing growth at all times but is obviously more laborious.

Biocides for industrial purposes need not have pharmaceutical grade purity, activities of 95% being acceptable for many purposes, provided that the impurity is not an embarrassment. Solids are more difficult to administer than liquids and both may require special handling precautions. Recent experiments using slowly leaching biocides in flow-through containers, thus maintaining a constant high inhibitory concentration irrespective of top-up, drag-out or infection level, have yielded promising results.

Choice of Biocide

There may appear to be a bewildering number of variables affecting biocide choice and a bewildering number of biocides available. In the writer's experience only about a dozen biocides are worthy of consideration, and although laboratory work is usually necessary before making the final choice, it is possible to narrow the field by asking the suppliers for pertinent data on the following lines:

1. Is the biocide effective at the pH of my system?
2. Is it inactivated by cationic/non-ionic/anionic emulsifiers?
3. Is it oil- or water-soluble or both?
4. Is it stable, volatile, hydrolysable etc.?
5. Is it a skin or lung irritant?
6. What is its spectrum of biological activity?
7. How is its activity affected by temperature?
8. Is it corrosive to steel, aluminium etc.?
9. Is it persistent?
10. Is it surface-active?
11. What is the cost of effective treatment?

The answers to these questions, if known, will be of great help, but for any given situation some may not be relevant. For example, lung irritation is of little import for a hydraulic oil treatment and cost is of less significance when the product protected is expensive.

In fact, apart from the industrial health implication, the whole field of infections in metal machining and deformation must be considered from an economic standpoint, and the cost of investigation and treatment weighed against foreseeable economies in production.

Although there is no substitute for experience in this field, it is possible to simplify the microbiology to a level which can be handled by any competent chemical or metallurgical laboratory. The following Appendix deals with taking and testing samples by people acquainted with the basic principles of microbiology as given in Chapter 1; the procedure was first published in *Tribology*, February 1969, and is reproduced with the Editor's kind permission.

Appendix

Microbiological Examination of Petroleum Products

Method 1. *Sampling Procedure*

STERILIZING

A convenient bottle for sampling is a screw-capped 28 ml Universal container. It can be sterilized by autoclaving but not by heating in a dry oven (the rubber cap liner is affected by oven treatment). A pressure cooker is a good substitute for an autoclave, 15 lbf/in^2 (121 °C) for 10 min being adequate. The cap is loosened slightly before sterilizing but must be screwed up again when the bottle has cooled. Alternatively, clean bottles can be filled with ethyl alcohol, shaken, allowed to stand for about 10 min, the alcohol poured off and the bottle sealed.

SAMPLING

When taking a sample the bottle is held in the left hand and the lid unscrewed with the right hand where it is retained whilst the sample is taken. Avoid touching the lip of the bottle or lid.

Emulsion samples are best taken from circulating emulsion at the sprays, as in stagnant tanks there is a tendency for organisms to rise with the cream or settle with the metal fines. If the sample has to be taken from a tap or valve allow the emulsion to run for about 30 seconds before taking the sample.

Straight oil tanks can be examined by taking a sample from the tank bottom drain or by appropriate dipping. A water-bottom, if it exists, is a useful sample. The higher the sampling position is in the tank, the fewer living organisms are likely to be present.

Fill the sample bottle to the shoulder. Note relevant information such as the time, the temperature, whether the machinery is running and the identity of the oil or emulsion.

STORAGE

If the period before testing is short the findings are likely to be more significant. Keeping samples too long at room temperature encourages excessive growth of anaerobic organisms. Samples can be stored in a domestic refrigerator at 2–4 °C. Do not deep freeze. Samples which must be sent for a considerable distance by post can be placed in a thermos flask filled with ice and inert packing.

TESTING

Thoroughly shake the sample before testing. Most tests will involve incubation, but cheap incubators can be made from pie warmers or even biscuit tins and light bulbs. Incubate at a temperature which is related to the temperature at which the sample has been used unless otherwise indicated, *e.g.*, 37–50 °C for rolling coolants, 30 °C for cutting oils, 25 °C for cold storage tanks. Work near a Bunsen flame as the up-current of warm air minimizes accidental contamination. If calibrated sterile pipettes are required they are best prepared by oven sterilization (160 °C for 3 h) in cans after lightly plugging the mouthpieces with non-absorbent cotton wool. Glass Petri dishes can similarly be oven-sterilized, or alternatively sterile disposable plastic Petri dishes can be obtained.

After testing, cultures can be disposed of by autoclaving or immersing in disinfectant overnight.

Method 2. Red-Spot Test

MATERIALS REQUIRED

5 mm diameter wire loop (24 SWG Nichrome wire) in holder, the plane of the loop to be at an angle of about 120° to the holder
A thin-walled metal tube, I.D. 9 mm (*e.g.*, a suitable cork borer)
Bunsen burner
Incubator at 30 °C or 37 °C; water-bath at 50 °C
Red-spot plates (Oxoid Ltd.) or ampoules (B.D.H. Ltd.)

Formalin (40% formaldehyde solution)
Bedacryl 122X (I.C.I. Ltd.) if permanent records are required.

PLATE METHOD

Allow a Red-spot agar plate to dry in the incubator for about ½ h with the dish slightly ajar and the agar surface facing downwards. Imprint one or more circles into the agar with the metal tube after holding the tube in the Bunsen flame for a few seconds to warm and sterilize it.

Inoculate each circle with one loopful of sample (approximately 0·01 ml). Sterilize the loop by flaming until it is cherry red before and after use. Try to fill the circle without overflowing (Fig. 5.12). Replace lid of Petri dish and allow to dry for about ½ h; then invert so that the agar surface faces downwards. Incubate overnight (16 h) or some other convenient period. 37 °C is a good general purpose temperature but 30 °C may be preferred for samples of cool origin.

Remove from the incubator and examine. Record the results in terms of red obliteration of the circle, which indicates microbial growth (Fig. 5.13); for example:

+ + + + +	complete red obliteration
+ + ±	half obliteration
–	no colour.

If a negative or slight positive is obtained it may be desirable to continue incubation and record the final result after 3 days. This is particularly so for hot rolling coolants which may contain slow-growing organisms. Alternatively a sample of hot origin can be tested in an ampoule (see below) and incubated in a water bath at 50 °C when it will frequently grow well and give a readable result overnight.

After the selected incubation period the result will continue to change even at room temperature and it is advisable to hold it by adding a few ml of formalin to the Petri dish lid thus exposing the organisms on the agar surface to the vapour. Several hours are necessary for a complete kill. Semipermanent records can be kept by flooding the agar, after exposure to formalin, with Bedacryl 122X and allowing to set.

Standards can be applied by the user of the test depending on circumstances. Similar coolant from similar sources with a comparable infection level will tend to give a consistent Red-spot result but if the method of use or formulation is different, results are not comparable. The test is of most value in rapid surveys, and for checking antibacterial measures. It will not detect anaerobes and gives very little response with yeasts and moulds.

AMPOULE METHOD

A full description of the ampoule modification is included in each kit. Briefly, the sterile ampoules containing Red-spot agar are snapped open and a sample measured in with a capillary. Ampoules can, if necessary, be incubated in the breast pocket, and can be preserved as previously described.

The agar used in the test has been fully described elsewhere[10] for those who would like to make their own from basic components.

Method 3. *Black-Spot Test*

MATERIALS REQUIRED

Wire loop as described for Red-spot test
Iron sulphite agar (Oxoid Ltd.)

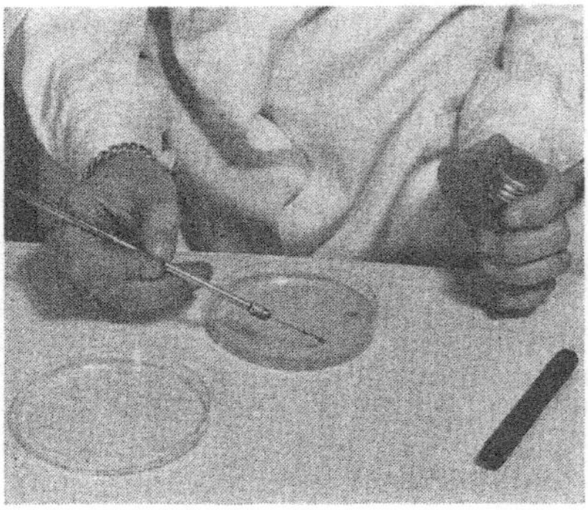

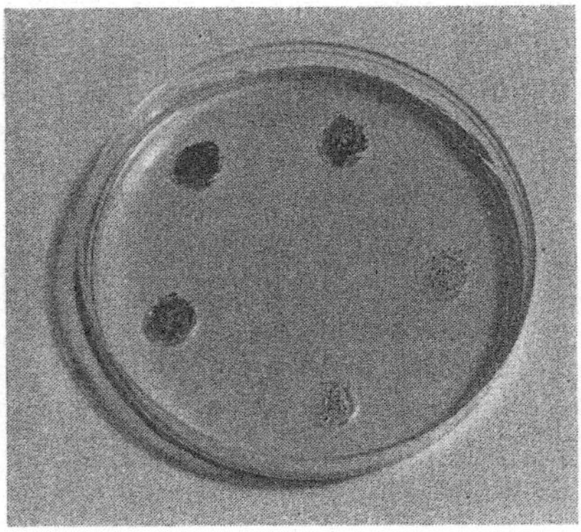

Fig. 5.11 *Laboratory simulation of emulsion in works use: three different emulsions each circulate from a stagnant reservoir to an aeration tower and back*

Fig. 5.12 *Transferring a loopful of sample to a "Red Spot" plate*

Fig. 5.13 *"Red Spot" plate test results after incubation*

Fig. 5.14 *Results of "Black Spot" tests:*
A *Negative*
B *Few anaerobic organisms – no H_2S generation*
C *Few H_2S-generating anaerobes*
D *Severe infection with H_2S generation*

Potassium dihydrogen orthophosphate
6 in. (150 mm) × $\frac{5}{8}$ in. (15 mm) rimless bacteriological test tubes
Rubber test tube seals (Astell Laboratory Service Ltd., Catford).

Dissolve 5 g potassium dihydrogen orthophosphate in a minimal amount of distilled water. Adjust pH to 7·0 with 5N NaOH, using an indicator such as bromothymol blue. Make up to 100 ml with distilled water. Autoclave 15 min at 15 lbf/in^2. Dissolve 23 g iron sulphite agar in 1 litre of water and dispense in 10 ml portions into bacteriological test tubes (with Astell seals) and sterilize by autoclaving for 15 min (see Method 1). Ensure that the seal is driven home before storing.

TEST

Melt sufficient tubes for the tests in hand by immersing them in a beaker of water and bringing to the boil. When completely melted, measure 1 ml sterile phosphate solution into each test tube using a sterile pipette and lightly flaming the neck of each tube and the phosphate bottle. Roll each tube between the hands to mix. Allow to cool until they can be comfortably held in the hand. At this temperature they should still be liquid. Alternatively transfer them to a water-bath at 50 °C.

Take a loopful of a sample using a well-flamed loop and plunge the filled loop into the molten medium. Flame the test tube mouth when the seal is removed, DO NOT PUT THE SEAL DOWN or finger the inner surface, and flame the test tube mouth again when the loop has been removed. Flame the loop. After firmly replacing the seal, roll the tube between the hands to mix. Incubate at 37 °C.

Record the results overnight and then at 24 h intervals for 5 days. Nodules of growth (colonies) will be seen in the depths of the medium. Some may produce sufficient gas to split the agar. If hydrogen sulphide is produced, black zones will appear around the colonies. Ignore surface growth. The method is only semiquantitative but in practice yields useful information. Very heavy anaerobic infections will usually be detected after overnight incubation, particularly if hydrogen sulphide is evolved. The black coloration can occasionally be phosphide precipitation rather than sulphide precipitation, but it has been shown using a large number of variants of the medium that a black coloration in the test is always associated with sulphide production ultimately (Fig. 5.14).

The following terminology may be helpful:

0–10 colonies = light anaerobic infection
10–30 colonies = moderate anaerobic infection
30–100 colonies = heavy anaerobic infection
More than 100 colonies = very heavy anaerobic infection.

The proportion of colonies producing hydrogen sulphide should also be recorded.

As a check of procedures an uninoculated tube can be incubated and should of course be negative.

Anaerobic microbes are often more sensitive to biocides than are aerobic organisms. Thus the disappearance of anaerobes but not aerobes after biocidal treatment usually indicates an inadequate biocide concentration.

It has recently been noted that some infected tubes, after prolonged standing in the light, become red. This is due to the slow growth of purple photosynthetic bacteria which must also have been present in the sample.

Method 4. *Detection of Fungi*

MATERIALS REQUIRED

Wire loop as described for Red-spot test
Corn meal agar (Oxoid Ltd.)
Beaker of alcohol
Spreaders (thin glass rods, about 6 in. (150 mm) long, with a right-angle bend about 1½ in. (40 mm) from one end).

METHOD

Dissolve 17 g corn meal agar in 1 litre water, dispense in convenient screw-capped bottles in 200 ml portions, and autoclave 15 min at 15 lbf/in². While still hot, dispense the contents of as many bottles as required into sterile Petri dishes flaming the bottle neck at each transfer. Cover the bottom (the smaller half) to a depth of about 5 mm. Leave the lid ajar for about 30 seconds to disperse steam and then cover fully and allow to set. Store at 4 °C if not required immediately. If necessary dry off before use as described in Method 2.

Store the spreaders with the short leg immersed in the beaker of alcohol.

Transfer a loopful of sample aseptically to the centre of a dish of medium. Take a spreader, holding the extremity of the long leg, flame off the alcohol and allow to cool for a few seconds. Use the short leg to spread the drop of sample evenly over the growth medium. Discard the spreader into the beaker of alcohol. Incubate the dish with the agar surface facing downwards at 25° or 30 °C. Examine after 2, 3 and 4 days. Fungi appear as fuzzy growths, often coloured green or black. Accurate identification is a job for the expert but the quantity of growth can be recorded as "light", "moderate" etc. The method is only semiquantitative.

The medium is not strictly selective and many bacteria will grow as smaller, adherent colonies. Alternatives to corn meal agar are Sabouraud's medium and malt extract agar.

Fungi are often an important component of the infecting microflora when the pH is low. Occasionally they become dominant at neutral or slightly alkaline pH after narrow-spectrum biocides have suppressed bacteria.

Method 5. *Detection of Yeasts* (Medium devised by J. W. T. Wimpenny)

MATERIALS REQUIRED

Tryptone soya broth (Oxoid Ltd.)
Agar no. 3 (Oxoid Ltd.)
Sodium chloride
Sucrose
Concentrated sulphuric acid
Johnson pH indicator paper no. 3651 (or equivalent)
Suitable sized conical flasks
Pipettes
Cotton wool
Sterile Petri dishes
Distilled water
5 mm Nichrome loop in holder
Beaker of alcohol
Spreaders.

TEST

The yeast agar medium is prepared in two parts and subsequently acidified to the correct pH as follows:

1. 15 g Tryptone soya broth, 10 g Oxoid agar no. 3 and 5 g sodium chloride are dissolved in 500 ml distilled water and autoclaved at 15 lbf/in² for 15 min, in a 1 litre conical flask with a cotton wool plug.

2. 100 g sucrose is dissolved in 500 ml distilled water and autoclaved at 10 lbf/in² for 15 min. Solution 2 is carefully and aseptically mixed with solution 1 so that no bubbles are trapped in the mixture. Both solutions must still be hot.

7 ml of a 20% (v/v) solution of Analar sulphuric acid in water is then added to give pH 3·6. Swirl carefully to mix. The pH may be checked using the indicator paper: a pH of 3·6 is indicated when the colour of the paper has *just* turned orange. Pour a sample into a watch glass for the check.

When the medium has cooled to about 50 °C it should be carefully poured into sterile Petri dishes. Each Petri dish should receive between 10 and 15 ml of the medium (*i.e.*, a depth of about 5 mm). The "plates" as they are called are now allowed to set and may be stored at 4 °C. If necessary dry off before use as described in Method 2.

The plates should now be inoculated using a Nichrome loop and a spreader with a sample of appropriate material as described in Method 4, and incubated at 30 °C with the agar surface facing downwards for 3–4 days.

The medium is reasonably selective if prepared properly although fungal colonies occasionally occur. The yeast colonies are recognizable as round, humped nodules, with a characteristic yeasty smell. The colour is usually creamy white but pink colonies are not unusual.

The method is semiquantitative; the amount of growth can be described in terms of light infection, heavy infection etc. Yeasts are usually a minor component of an infection unless the pH is low or a narrow-spectrum biocide has been used against bacteria. Yeast infections can be exceptionally severe if unchecked. Under the microscope the yeast cells are often ovoid, frequently with a daughter cell budding off, but filamentous forms are not uncommon.

Method 6. *Quantitative Estimation of Aerobic Bacteria*

MATERIALS REQUIRED

5 mm wire loop in holder
Tryptose blood agar base (Oxoid Ltd.)
Beaker of alcohol
Spreaders
Bacteriological test tubes with Astell seals containing 1 ml distilled water, sterilized by autoclaving for 10 min at 15 lbf/in².

TEST

Accurate quantitative estimations of bacteria will be beyond the expertise of most industrial laboratories, but if required a method is given in Ref. 10.

A very good indication of numbers will be given by following the procedure described in Method 4 but using Tryptose blood agar base (Oxoid Ltd.) at 30 g/litre instead of corn meal agar. The numbers of bacterial colonies arising after incubation can be counted if this is not more than about 500. Incubation is best carried out at 30 °C for cutting and grinding fluids and 37 °C for rolling emulsions or other fluids normally maintained at an elevated temperature. Three days' incubation is usually adequate although some slow-growing organisms will not develop into a visible colony until about the fifth day. The method requires that individual living microbial cells be allowed to reproduce into a visible colony, and hence by counting developed colonies, the number of original cells can be arrived at. For many technical reasons this must always be considered a minimum number.

Assuming that the loop delivers about 1/100 ml and that not more than

500 colonies can be counted on a plate, a single plate prepared as described could be used for infections of up to 50 000 organisms per ml, a very modest level in most industrial situations. It is therefore advisable to prepare dilutions in distilled water by transferring a loopful of the sample to a test tube containing 1 ml sterile distilled water (flaming the tube neck on opening and before sealing) and after mixing well, preferably mechanically, flame the loop and transfer a loopful of this dilution to a further sterile tube of distilled water which is also mixed. These dilutions, labelled 10^{-2} and 10^{-4} respectively, are then tested as for the original sample.

After incubation the three plates are examined. A typical result might be

Original sample	Uncountable
1/100 ml of 10^{-2} dilution	290 colonies
1/100 ml of 10^{-4} dilution	3 colonies.

Best answer = 290×10^{-4} = 2 900 000 aerobic bacteria/ml.

The best information is derived from a plate having between 30 and 300 colonies if one is available.

Using other growth media, selective cultures may be made of organisms causing common sepsis but it is strongly advised that these tests should be left to professionally qualified microbiologists. It should not be forgotten that all *laboratory* cultures are potentially dangerous, even those in the media listed in this article, and they should all be treated with respect and disposed of as previously described.

Method 7. Microscopy

The following notes are intended for those who have access to, or intend to purchase, a microscope. It is not possible to teach microscopy in a brief article, but most microscope suppliers cover basic points in their handbooks.

The ideal optical set-up is a phase-contrast transmitted-light microscope with at least low and high power objectives. The latter should be $\times$ 40 and although an achromatic objective is often adequate, a fluorite objective will probably justify the extra cost. The eyepiece should be $\times$ 10 compensating, although a $\times$ 10 Huyghenian is a cheaper alternative.

With the phase-contrast microscope, samples are examined wet with a cover-slip over them. This should be 0 or 1 gauge in thickness unless contra-indicated by the microscope manufacturers.

Interpretation of specimens comes with long experience, but indications of microbial infection are particles exhibiting genuine motility, as distinct from Brownian (vibratory) movement, and regularity of shape and size. Because of the magnification involved even a few microbes per microscope field represent several million per ml.

When examining emulsions microscopically it is often impossible positively to identify micro-organisms amidst the oil droplets and inorganic debris. However, continued experience of the same emulsion leads to a good knowledge of the normal oil droplet size distribution; and as degradation of emulsifiers by microbes alters this distribution, indirect evidence of infection is obtained. Droplet size can be accurately measured using the image splitting eyepiece (Vickers Instruments Ltd., York).

General Note

The methods described in this Appendix work well for water-miscible samples such as oil emulsions and synthetic coolants, and moderately well for infected straight oils. Straight oils can be investigated by examining free condensate water in the system on the assumption that unless this is infected there is unlikely to be a microbiological problem. If there is no obvious condensate in the system an oil sample can be vigorously shaken with an equal volume of sterile water in the laboratory, and the aqueous phase tested after settling. Alternatively the straight oil can be deliberately made into an emulsion by adding a non-toxic emulsifier such as "Corexit" (Esso Chemicals Ltd.).

Obviously the dilution method in Method 6 cannot be used unless the sample is water-miscible.

* * *

Figs. 5.2, 5.7, and 5.10 are reproduced by kind permission of the Editor, *New Scientist.*

References

[1]HILL, E. C. AND GIBBON, O. M. '*Corrosiveness of emulsions*'. Metal Finishing, **15**, 395–398 (1969).

[2]BOURNE, L. B. '*Industrial dermatitis*'. British Journal of Hospital Medicine, 196–208 (November 1968).

[3]HILL, E. C. '*The significance and control of micro-organisms in rolling oils and emulsions*'. Metals and Materials, **1**, 294–297 (1967).

[4]CHEMISTRY RESEARCH 1956 (*Annual Report of the Chemical Research Laboratory*). p. 61. Her Majesty's Stationery Office, London (1957).

[5]HILL, E. C. AND PENBERTHY, I. O. '*Microbiology of steel mills and its relation to sheet corrosion*'. Metals and Materials, **2**, 359–366 (1968).

[6]HILL, E. C. '*Microbiology of hydraulic systems*'. Fluid Power International, 56–57 (September 1968).

[7]HILL, E. C. '*Designing out microbial infections*'. Engineering Materials and Design, **10**, 1924–1925 (1967).

[8]ROSSMORE, H. AND BRAZIN, J. G. '*Control of cutting oil deterioration with gamma radiation*'. In '*Biodeterioration of Materials*' (ed. by A. H. Walters and J. J. Elphick), pp. 386–402. Elsevier Publishing Co. Ltd., London (1968).

[9]CALLELY, A. G. '*Biocides in industry*'. Process Biochemistry, **3** (11), 13–16 (November 1967).

[10]HILL, E. C., DAVIES, I., PRITCHARD, J. A. V. AND BYROM, D. '*The estimation of micro-organisms in petroleum products*'. Journal of the Institute of Petroleum, **53**, 275–279 (1967).

CHAPTER SIX: J. J. ELPHICK

Microbial Corrosion in Aircraft Fuel Systems

Introduction

The majority of cases of microbial deterioration of industrial materials and equipment can be successfully controlled by one or other of the standard preventive techniques. A material which is susceptible to enzymic degradation may be replaced by a resistant one; a biocide may be applied or a sterilization process used to prevent microbial development; the environment may be modified so as to exclude the possibility of active growth. Fortunately the changes in design and production necessitated by such measures are in most instances relatively inexpensive and easy to introduce and so present no major hardship either to the manufacturer or to the user. Occasionally, however, particularly when a microbiological problem affects a complex industrial product or process which must be constructed, and function, as economically as possible, there may be greater difficulty in finding a solution which will satisfy all the technical and financial requirements of manufacturer and operator.

This state of affairs is illustrated by the problems of corrosion and blockage caused by microbial growth in aircraft fuel systems and associated ground installations. Although a number of control measures are theoretically available, all have certain disadvantages which render them unpalatable to one or more of the four major industries concerned – the airframe manufacturers, the aero-engine manufacturers, the oil companies and the aircraft operators. This chapter attempts to illustrate the practical difficulty of protecting a technically complex and financially vulnerable product from what is fundamentally a fairly simple microbial corrosion problem. It will become apparent that although the various ameliorative measures adopted to date have brought about some improvement in the situation, further research is needed if fully effective control is to be achieved.

Origins of the Problem in Aircraft

Serious microbiological problems in aircraft were virtually unknown until well into the 1950's, when it was noted that cases of corrosion in aircraft fuel tanks in Australia and the U.S.A. were accompanied by the presence of varying amounts of free water which contained a dark sludge. This sludge proved on examination to consist largely of masses of fungal growth, frequently accompanied by numerous bacteria. At first there was a tendency to regard this association of micro-organisms with fuel tank corrosion as fortuitous, and the corrosion process was attributed to such well-known phenomena as accumulation of inorganic ions in the residual water. However, as further cases were reported it became clear that microbial sludge was invariably present in the affected fuel tanks and a more positive involvement was thus indicated. Although it was to be several years before microbial corrosion of structural alloys was conclusively demonstrated in the laboratory, it was soon widely accepted that fungi and bacteria were in some way responsible for the problem. In addition, it was found that masses of fungal growth became detached from the tank floor and caused blockage of fuel system filters and malfunctioning of capacitance gauges. Moreover, it soon became apparent that the problem was a widespread one. Although the most rapid growth and corrosion occurred in aircraft operating in tropical and subtropical areas throughout the world, cases of this kind were occasionally found in temperate zones. Since the early 1960's the industries concerned have become increasingly aware of the seriousness of the situation, and considerable effort has been directed towards the elimination of microbial growth from storage and supply systems both in aircraft and on the ground.

Why was it that after nearly fifty years of aviation development, aircraft should suddenly become afflicted with this new microbiological hazard? The standards of aircraft maintenance and inspection were such that it could not have previously gone unnoticed in this serious form. The answer lies in the interaction of two advances in aircraft design which were being introduced during that period. The first was the entry into world-wide service of gas-turbine engined aircraft, which used kerosene rather than gasoline fuel. Kerosene fuels, as will be seen below, are an ideal food source for that group of hydrocarbon-utilizing microbes best adapted to grow in aircraft fuel systems. The second factor, of particular relevance to the corrosion problem, was the increasing adoption of the integral or "wet wing" type of aircraft fuel tank. Earlier aircraft had carried fuel in rubberized fabric containers ("bag tanks") situated in spaces in the structural framework of the wing. Modern aircraft wings are of box-like construction; the upper and lower surfaces are main load-bearing members and are constructed of machined alloy panels. The presence of this substantial and readily sealable box, or series of boxes, in the wing permitted the direct use of the space within for carrying fuel, and removed the necessity for bag tanks.

Occurrence of Micro-Organisms in Fuel

The remarkable efficiency of the reproductive and disseminating mechanisms of the fungi and bacteria ensures that the soil, air and natural water throughout the world contain vast numbers of their spores and other regenerative bodies. It is inevitable that any substance exposed to the natural environment will accumulate contaminating microbes of many different species, as is made clear in Chapter 1. Hydrocarbon fuels are no exception to this rule: the normal displacement of air as bulk storage tanks are filled and emptied will bring airborne and dustborne spores into intimate contact with the fuel. The seepage of ground water into storage facilities will occasionally provide an additional source of infection. It must therefore be accepted that in normal operation, bulk fuel supplies are likely to become contaminated with significant amounts of microbial matter during their passage from the refinery to the final dispensing installation. The same fact applies to aircraft fuel tanks: they too are subject to contamination by airborne micro-organisms, and although the air displacement will be on a much smaller scale the possibility of direct infection of aircraft must not be overlooked.

While the viability of bacterial cells in fuel is generally low, and many species remain alive for only a few days or even hours, the spores of most fungi will survive suspension in kerosene for many months. Viable material will of course resume active growth and reproduction when it is returned to favourable nutrient and environmental conditions. The great majority of species entering fuel supplies never find these conditions fulfilled, for although the water and mineral salts necessary for growth are frequently available, they are unable to utilize fuel as a source of carbon. Their importance in fuel microbiology therefore lies solely in their tendency to appear as contaminants during cultural investigations, where they may lead to confusion unless appropriate techniques and selective media are employed to suppress them. The problem in aircraft fuel systems is caused by the comparatively few species which can derive the carbon they need for growth by degrading various constituents of the fuel. Each of these organisms produces the enzymes necessary for the oxidation or hydroxylation of a particular range of hydrocarbons. For example, the aliphatic hydrocarbons which are the most important components of aviation fuels are degraded by a progressive terminal oxidation process brought about by oxygenase enzymes (enzymes catalysing a direct reaction of their substrate with molecular oxygen).

The ability of the fuel to serve as a carbon source for microbial metabolism is indeed the key factor in fuel microbiology. Experience has shown that the C_9 to C_{16} alkanes of aviation fuels are particularly susceptible to degradation by that group of hydrocarbon-utilizing organisms best adapted to grow in the environment of ground storage and aircraft fuel tanks. It has also been found that the growth rates of any one of these micro-organisms in the various types of aviation turbine fuel are very similar, and that diesel fuel provides an equally acceptable food source. Gasolines, on the other hand, are not readily degraded by these organisms and this accounts for the failure of the problem in question to reach significant levels in piston-engined aircraft. At one time tetraethyl lead (TEL), widely used as an anti-knock additive in aviation gasolines, was suspected to act as an antimicrobial agent, but the work of Hazzard[1] on the growth of fungi in kerosene/gasoline mixtures suggests that TEL has very little inhibitory effect and that the comparative resistance of gasoline is in fact due to its hydrocarbon constitution.

Early work on general fuel microbiology tended to concentrate on the bacteriology of the material, and it was found that various *Pseudomonas* species were the major hydrocarbon-degrading organisms associated with fuels at normal ambient temperatures. Pseudomonads, usually referred to the species *P. aeruginosa*, are frequently isolated from water phases associated with aircraft fuel systems in spite of their limited viability in fuel. However, the predominant organism present in aircraft tanks is undoubtedly the fungus *Cladosporium resinae*, sometimes incorrectly referred to as *Hormodendrum resinae* (see Fig. 6.1). This observation has been consistently confirmed by

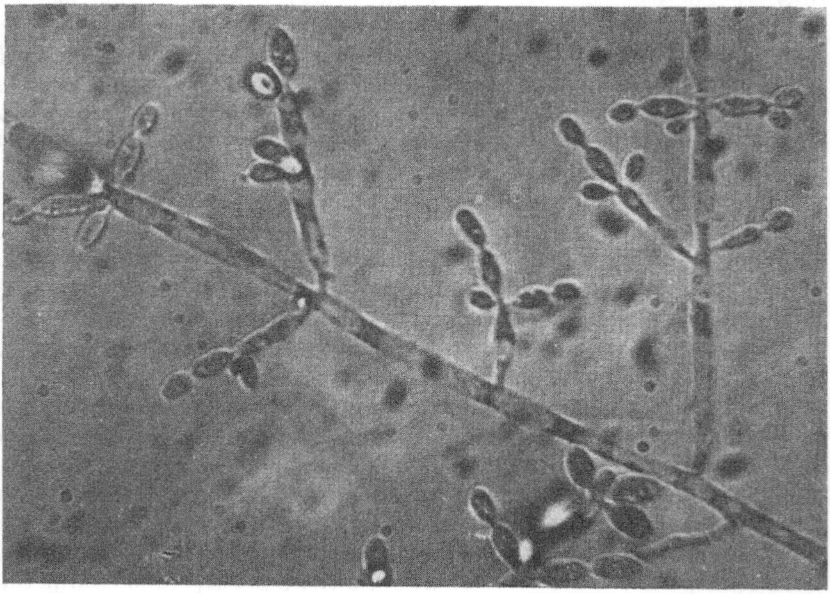

Fig. 6.1 Cladosporium resinae: *photomicrograph of sporing structure* (× 300). (*Crown Copyright reserved*)

almost all workers who have investigated the problem. Darby, Simmons and Wiley,[2] in the most authoritative review to date on the mycology of fuels, found that in addition to *C. resinae*, *Alternaria tenuis*, *Paecilomyces varioti* and *Fusarium* spp. are also capable of "very active" utilization of fuel and list 17 other species with moderate or slight activity. Nevertheless, only rarely do these fungi compete successfully with *C. resinae* in the aircraft environment. Several species of yeast, *e.g.*, *Candida lipolytica*, have occasionally been isolated from aircraft and can be shown to utilize aviation kerosenes. Apart from *Pseudomonas* and related bacteria, it has often been suspected that sulphate-reducing bacteria (see Chapter 3) may be involved in the corrosion of aircraft fuel systems. Although a number of workers record failure to detect them, Iverson[3] reports their isolation from corrosion pits in integral tanks. Corrosion of aircraft engine fuel pumps has occasionally been reported, but this problem has usually been traced to the growth of sulphate-reducing bacteria in ground fuel storage tanks. The hydrogen sulphide so produced dissolves in the fuel and is liberated as free gas as the fuel temperature rises on entering the engine fuel system.

Most of the species of fungi and bacteria isolated from fuel are common and fairly "omnivorous" saprophytes which, although able to degrade hydrocarbons, frequently exist on alternative carbon sources in nature. Their occurrence in fuel systems throughout the world is a natural consequence of their ubiquity. It is of interest to note that *C. resinae* is more restricted in its range of natural habitats and is less commonly encountered by routine methods of examination of the atmosphere, soils etc. Its original natural habitat is thought to be resinous conifer timber, but because of its ability to metabolize creosote it occurs fairly frequently on timber which has been impregnated with that preservative – telegraph poles, fencing, railway sleepers etc. It is probable that airborne spores from such sources are responsible for much contamination of bulk fuel stocks; Harvey[4] has shown that *C. resinae* may form a small but significant part of the aerial spore population. In addition, Parberry[5] has demonstrated that this species can regularly be isolated from soil in the neighbourhood of creosoted wood, which lends at least some support to the theory that its spores may be carried into ground storage tanks in water seepage.

Environmental Factors and Growth

Temperature and the availability of water are the principal factors determining whether, and where, microbial development will occur in ground storage installations and aircraft fuel systems. Other factors – oxygen and carbon dioxide concentrations (both gases are readily soluble in fuel) and pH – are seldom, if ever, growth-limiting in practice. Aviation kerosenes may contain up to 100 p.p.m. of dissolved water, depending on temperature, but although micro-organisms may remain viable in dry or water-saturated fuel, active growth will not occur. It has been demonstrated (Elphick[6]) that *C. resinae* can grow on kerosene-soaked surfaces in air when the ambient relative humidity is in excess of 80%. Nevertheless, in both ground installations and aircraft, growth is seldom found elsewhere than in contact with deposits of free water, and the location and extent of growth is determined by the pattern of water deposition and retention.

Apart from ground water seepage (probably an infrequent occurrence), and the accidental entry of rainwater through open hatches (also hopefully rare), the main source of water deposition in bulk storage tanks is likely to be condensation in the air space resulting from diurnal temperature fluctuations. In the aircraft wing, water dissolved in the fuel at ground temperature is precipitated from solution at the sub-zero temperatures reached during high altitude flight. It is possible that particulate matter such as fungal spores present in the fuel will serve as foci for the formation of water droplets – perhaps a significant factor in assisting microbial contamination of water deposits. However, the greater part of the water present in integral tanks arises, as in the case of ground storage tanks, by condensation. Humid air drawn into the partly filled tanks during descent to ground level forms heavy condensation, frequently in the form of ice, on the cold surfaces in the air spaces of the tanks. A third method of entry, the accidental uptake of water from ground installations during refuelling, can only occur if there are serious deficiencies in the fuel supply system, *e.g.*, a broken water separator, and is unlikely to occur at a major airport.

Fungal growth in water deposits occurs first at the interface with the fuel, and frequently the mycelium forms an envelope around a water droplet which tends to anchor it in position. As growth continues, the fungus will tend to fill the water phase and adhere to the tank lining. Further increments of water will be unable to dislodge an established fungal colony, and may serve to dilute and wash away metabolic waste products which would otherwise inhibit growth by "staling". It is thus possible to consider this situation as an efficient continuous culture system, with regular aeration, water and mineral salt replacement, and abundant available carbon. One of the most dangerous aspects of interfacial growth by *C. resinae* is its ability to produce sporing structures in the kerosene layer almost as soon as a clearly visible colony has been established. This results in a rapid build-up of high spore concentrations as disturbance on emptying and refilling liberates these spores into the fuel.

Until recently it was held in some quarters that the solution to the problem of microbial growth in aircraft lay in the elimination of water by proper drainage procedures and the correct design of integral tanks to eliminate undrainable areas. Unfortunately, experience has shown that although the retention of water can be reduced by these measures it is impossible to drain the many droplets and small puddles which adhere to the tank floor. Because of the low dihedral angle of a modern aircraft wing these droplets are unaffected by gravitational flow, and even resist the currents caused by fuelling and defuelling. It is imperative, then, to recognize that water is an unfortunate but ever-present contaminant of integral wing tanks.

Whereas the pattern of retention of water dictates the site of growth in a fuel tank, the rate of microbial development is determined by temperature. The organisms comprising the microflora of storage and aircraft tanks exhibit normal growth/temperature curves. *Cladosporium resinae*, for example, grows very slowly at 10 °C and its rate of development increases steadily to a maximum at 30–35 °C, then drops quite sharply to become negligible again at about 45 °C. This fairly high optimum growth temperature substantiates the generalization that the problem is predominantly a tropical one, while as slower but significant growth occurs in the 15–20 °C range its presence in temperate zones is also explained.

Although ambient temperature determines the rate of growth in ground installations, it does not directly define the rate of development in aircraft. Here it is substantially modified by a factor of even greater importance, namely the effect of flight pattern on actual fuel tank temperatures. The fuel in an aircraft in flight on a long distance route may reach temperatures as low as −40 °C, and even on short haul routes it may fall to −20 °C. On landing, the wing structure and the residual fuel it contains will take a considerable time to rise to ambient temperatures; indeed, ice may persist in the integral tank of a transatlantic airliner for up to 24 hours after landing at London. This factor implies that conditions inside a tank may be conducive to microbial growth for surprisingly limited periods, and undoubtedly explains the comparative freedom of many commercial aircraft from microbial troubles. Naturally, higher tropical ambient temperatures and the much more intense solar radiation encountered in these areas will result in more rapid heating of the aircraft wing. Nevertheless, time spent on the ground is by far the most important factor deciding whether or not an aircraft will suffer the expensive consequences of microbial corrosion of its fuel system.

The Corrosion Process

Two principal mechanisms have been postulated by which micro-organisms may induce corrosion of aluminium alloys employed in the wing structure. Firstly, fungi and bacteria produce a wide variety of water-soluble organic acids as waste products of their metabolism, and these are secreted into the environment. Indeed, water drained from infected tanks frequently gives pH readings of between 3 and 4, and one can envisage even more acid conditions existing immediately beneath a fungus colony. Hence a purely acid corrosion of structural metal may occur, as has been demonstrated by Blanchard and Goucher[7] and Hedrick *et al.*[8] An alternative theory, which possibly commands more widespread support, is based on the fact that little or no oxygen can reach the lower layers of an actively growing microbial colony, as explained in Chapter 1. The gradient so formed will establish an oxygen concentration cell of an adequate potential difference to lead to electrochemical corrosion. Solution of aluminium ions beneath the colony will result in pitting and their combination with hydroxyl ions will lead to the formation of tubercles of oxidation products. An account of this mechanism has been given by Iverson,[3] who further suggests that under conditions of severe oxygen depletion sulphate-reducing bacteria may develop in the corrosion pits. These may then engender further corrosion by the mechanism suggested by Booth and Tiller.[9] It is interesting to speculate that a microbial community in which a growth of *Cladosporium resinae* overlies microaerophilic bacteria, such as the pseudomonads described by Traxler and Bernard,[10] would produce even steeper oxygen concentration gradients than would a colony of *C. resinae* on its own.

Some typical examples of the growth of *C. resinae* in aircraft integral fuel tanks are illustrated in Figs. 6.2 and 6.3.

Control of the Problem

It follows from the above that fuel and water should preferably not come into contact with a bare metal surface, and indeed in present-day aircraft the integral tanks are seldom unprotected. Joints, fastenings etc. are sealed with a suitable material, usually polysulphide rubber, while the entire inner tank surface is coated with a fuel-resistant material, in most instances nitrile rubber or polyurethane. When microbial corrosion was first recognized it was thought that microbes made contact with the metal through imperfections ("holidays") in the coating material. However, the ineffectiveness of multiple coat systems in eliminating corrosion lends credence to the commonly held view that both nitrile rubber and polyurethane are penetrated by fungal growth. The work of Hedrick and Gilmartin[11] and others emphasizes the need for closer study of the initiation of microbial corrosion through fuel tank coatings.

As stated in the Introduction, the successful control of microbial growth and its unwanted effects in aircraft fuel systems has been complicated by the necessary but conflicting technical and economic requirements of the industries concerned. A brief survey of some of the theoretical possibilities will illustrate the difficulty of achieving satisfactory prevention in practice.

Protection of the Aircraft Structure

The need to facilitate the drainage of water from the aircraft wing is now fully established as a necessary design feature, and most aircraft in service today have been built to exclude any undrainable areas from the tanks. Water drains are designed and sited to prevent large puddles forming, attention is paid to the through drainage of water between the stringers to the drain points, and areas where machining of the bottom wing planks may produce low spots are filled with light weight void filler. Further modification of wing design is unlikely to produce a significant improvement in the situation.

The major area where technical advance in construction may be beneficial is the development of improved fuel tank coatings. Clearly, factors of adhesion, resistance to fuel, flexibility over the operating temperature range etc. must be paramount, but low water absorption and resistance to fungal penetration are other properties which would be extremely valuable in a tank coating. For example, a material which will resist the adhesion of water droplets might usefully delay the establishment of firmly attached fungal

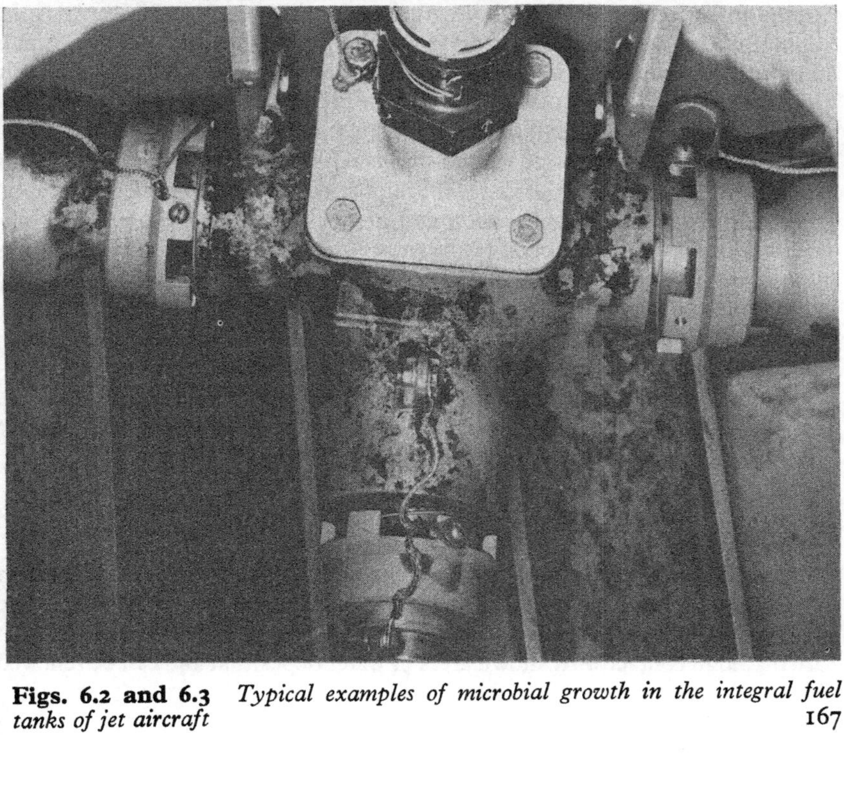

Figs. 6.2 and 6.3 *Typical examples of microbial growth in the integral fuel tanks of jet aircraft*

colonies. Much thought has been given to the incorporation of biocides in tank coatings. To be practicable, such a biocide would have to be fully compatible with the coating material, virtually insoluble in fuel, slightly soluble in water (to withstand continuous leaching and give an adequate service life), and yet be sufficiently active in the aqueous phase to inhibit microbial growth in volumes of several millilitres of water. Because of the weight penalty it is undesirable to increase the thickness of the protective coating above the present 1–1·5 thousandths of an inch. It is perhaps hardly surprising that so far no biocide has emerged that will produce sufficiently sustained antimicrobial activity from the loadings obtainable in this thickness of material. It is possible that biocide-containing coatings may delay fungal penetration and hence corrosion (although current evidence does not altogether support this) even though activity is not sufficient to prevent growth in the water phase. However, the parallel problems of gauge failure and filter blockage demand that growth be completely eliminated, rather than just corrosion prevented.

Fuel Cleanliness

The view that the oil companies should eliminate fungal contamination from the fuel they supply is at the moment as unrealistic as the demand that airframe manufacturers should build fuel tanks that retain no water. All the major oil companies are conversant with the problem and take conscientious measures to reduce the amount of water present in their storage and supply installations, and to prevent its entry into the aircraft. Bearing in mind the vast bulks of fuel handled, it is equally impossible to prevent its occasional contamination by spores of *Cladosporium resinae* and other microbes. Such improvements are of course technically feasible; more efficient filtration methods to remove fungal spores, and perhaps even fuel sterilization techniques, could be developed. However, the introduction of such measures would inevitably lead to increased fuel costs, which would be particularly unwelcome to operators whose flight routes happen to keep their aircraft free from trouble. The supply of two grades of fuel, one "cleaner" than the other, would present the companies with enormous distribution and storage difficulties and would result in even higher fuel prices. One is forced to the conclusion that at the present time the supply of substantially fungus-free fuel at an economic price is impossible.

Fuel Biocides

The application of biocides to fuel is currently the most important weapon in the fight against microbial contamination of aircraft fuel systems. One method is to employ cartridges of water-soluble, fuel-insoluble biocide which are placed in contact with known areas of water deposition and can be renewed when exhausted. Strontium chromate, already known as an anticorrosive compound, can be used in this way. An obvious difficulty is to ensure that all the water in the tank comes into contact with the biocide, and although sometimes reported as effective such cartridges have found only limited use.

The more usual method is to add a hydrocarbon-soluble biocide to the fuel. The two such materials at present in general use have high partition coefficients with respect to water, and thus low concentrations in the fuel will migrate to build up higher, toxic concentrations in any water phase in contact with the fuel. The first material to be employed in this way was ethylene glycol monomethyl ether (EGME; methyl cellosolve), which was already employed as an anti-icing additive to aircraft fuels. It is not a highly active substance and under severe operating conditions is effective only if applied in every fuel uplift. Because of this and the relatively high concentration required in the fuel (0·1–0·15%), it is expensive to use and hence not highly favoured by commercial airlines. It is however used by a number of military air forces throughout the world.

The second material is a proprietary mixture of organoborinanes – 2,2′-oxybis(4,4,6-trimethyl-1,3,2-dioxaborinane) and 2,2′-(methyltrimethylene-dioxy)bis-(4-methyl-1,3,2-dioxaborinane). This formulation, used at 150–270 p.p.m. in the fuel, is more active than is EGME and may be used as an intermittent treatment for fuel tanks. The various factors which should be taken into account when treating aircraft with intermittent doses of biocide have been examined by Elphick and Hunter.[12]

The use of biocides in fuel naturally arouses the interest of the aero-engine manufacturers, who are rightly somewhat wary of possible turbine damage by combustion products of these materials. EGME, being a carbon-hydrogen-oxygen compound, is regarded as safe but the more effective organoboron formulation has met with some suspicion. It was originally suggested that it should be used as a fill-and-drain treatment while an aircraft was standing on the ground (*e.g.*, during maintenance), a procedure which left the operator with a large volume of fuel he could not use for its intended purpose because of its biocide content. However, the occasional burn-off of treated fuel has now been approved for most engines, and this has done much to establish the use of organoboron as an intermittent biocide. Intermittent treatments are carried out during maintenance periods so as to allow maximum contact time between biocide and water phase. Severe leaching of biocide from the fuel will occur once the regime of water condensation on landing is renewed. The required frequency of application is still difficult to determine, and an operator must gain experience of his needs by trial and error. Much more basic information on the ecology and rates of growth under various aircraft conditions is necessary before an accurate forecast of treatment requirements can be attempted. The search for improved fuel biocides is continuing; Rogers and Kaplan[13] have screened 97 possible candidates and other workers are active in this field. However, the need for extensive engine trials of new materials means that the development of, and granting of approval to, a new product can be a long and expensive process. It is to be hoped that this will not diminish the enthusiasm of the chemical industry to seek and test such compounds.

Microbial Growth in Supersonic Aircraft

The current development of supersonic transport aircraft by various countries raises the interesting question of their possible susceptibility to microbial problems. The danger of corrosion will of course be greater in an aircraft constructed from conventional alloys rather than stainless steel (which will be used for the faster aircraft), but the danger of blockage and gauge failure will be present irrespective of the metal employed. There will be fundamental differences between the temperature and moisture conditions prevailing in supersonic aircraft fuel tanks and those of subsonic aircraft. Whereas those of a subsonic plane cool down to the air temperatures prevailing at altitude, the tanks of a supersonic transport will tend to heat up as a result of the greatly increased air friction encountered at speeds faster than that of sound. Clearly, if this heating effect is great enough, fuel temperatures may rise sufficiently to sterilize the fuel. On the other hand, maximum fuel temperatures would be expected to be maintained for only one or two hours during, say, a supersonic transatlantic flight, and if they are insufficiently high to achieve sterilization, fuel tank temperatures will be favourable for growth for much longer periods than occur in subsonic aircraft. The absence of the growth-inhibiting cold periods experienced in conventional aircraft may indeed be critical.

However, the absence of very cold surfaces will greatly reduce, or even eliminate, the accumulation of condensed water on landing, and the ability of supersonic transports to avoid serious microbial contamination will probably rest on whether or not they suffer significant water deposition. Should water persist in their fuel tanks, the effect of temperature changes on the microbial population will become significant. Prolonged high temperature may inhibit the growth of some organisms, such as *Cladosporium resinae*, which are now prevalent in fuel systems, but permit the development of thermophilic species. These may or may not have the same ability to adhere to the tank floor and cause corrosion.

Although it is interesting to speculate on the problem, it is impossible accurately to forecast the microbiological status of the fuel systems of supersonic transports. Until further development and flight trials have established the relevant temperature and moisture conditions, it is profitless to theorize further.

Conclusions

Microbial growth and corrosion in the fuel systems of aircraft has become an important world-wide problem which is still prevalent and which will be difficult to eradicate completely. Many practical questions remain unanswered. Is it possible further to reduce the risk of microbial infection of fuels without greatly increasing their cost? Can improved biocidal or fungus-resistant fuel tank coatings be developed? Can more effective fuel-soluble biocides be found? Would a deeper understanding of the relationships between growth and the fuel tank environment lead to more economic use of control methods? Until further research can provide at least some solutions, truly satisfactory containment of the problem will not be achieved. All the industries concerned have useful contributions still to make, and the closer the co-operation between them the more rapidly will progress be made.

* * *

References

[1]HAZZARD, G. F. '*Fungal growths in aviation fuel systems: Part 4. Fungi in aviation fuel systems in Australia and the Far East*'. Report 252, Defence Standards Laboratories, Australia (1963).

[2]DARBY, R. T., SIMMONS, E. G., AND WILEY, B. J. '*A survey of fungi in military aircraft fuel supply systems*'. International Biodeterioration Bulletin, **4**, 39–42 (1968).

[3]IVERSON, W. P. '*A possible role for sulphate reducers in the corrosion of aluminium alloys*'. Electrochemical Technology, **5**, 77–79 (1967).

[4]HARVEY, R. '*Air spore studies at Cardiff*'. Transactions of the British Mycological Society, **50**, 479–495 (1967).

[5]PARBERRY, D. G. '*The soil as a source of* Cladosporium resinae'. In *Biodeterioration of Materials*, ed. by WALTERS, A. H. AND ELPHICK, J. J., 371–380. London: Elsevier Publishing Co., 1968.

[6]ELPHICK, J. J. Appendix II to *An Analysis of Certain Microbiological Matter in Aircraft Fuels*, Shell International Petroleum Co. Ltd., London (1961).

[7]BLANCHARD, G. C. AND GOUCHER, C. R. '*Corrosion of aluminium by microbial cultures*'. Developments in Industrial Microbiology, **6**, 95–104 (1965).

[8]HEDRICK, H. G., ALLER, C. E., HALKIAS, J. E., HILDEBRAND, J. F., AND GILMARTIN, J. N. '*A current evaluation of a microbiological corrosion system for studying the effects on structural aluminium alloys*'. Developments in Industrial Microbiology, **6**, 117–123 (1965).

[9]BOOTH, G. H. AND TILLER, A. K. '*Anaerobic corrosion of aluminium by sulphate-reducing bacteria*'. Corrosion Sci., **8**, 549–555 (1967).

[10]TRAXLER, R. W. AND BERNARD, J. M. '*The utilization of* n-*alkanes by* Pseudomonas aeruginosa *under conditions of anaerobiosis. Part I. Preliminary observations*'. International Biodeterioration Bulletin, **5**, 21–25 (1969).

[11]HEDRICK, H. G. AND GILMARTIN, J. N. '*A detection study of microbiological penetration of aircraft fuel tank coatings*'. Developments in Industrial Microbiology, **6**, 124–132 (1965).

[12]ELPHICK, J. J. AND HUNTER, S. K. P. '*Evaluating biocidal fuel additives for intermittent use in aircraft fuel systems*'. In *Biodeterioration of Materials*, ed. by WALTERS, A. H. AND ELPHICK, J. J., 364–370. London: Elsevier Publishing Co., 1968.

[13]ROGERS, M. R. AND KAPLAN, A. M. '*Screening prospective biocides for hydrocarbon fuels*'. Developments in Industrial Microbiology, **9**, 448–467 (1968).

CHAPTER SEVEN: N. W. LE ROUX

Mineral Attack by Microbiological Processes

Introduction

We are now going to leave the field of corrosion, and examine aspects of microbiology which should be of interest to extractive metallurgists: some aspects of microbial attack, either direct or indirect, on minerals. I will only touch very lightly upon the question of economics and economic feasibility; and large-scale applications are dealt with by A. W. Fletcher in the final chapter. The processes that will be considered here can be grouped under the general heading of *geomicrobiology*, that is, the association of geology and microbiology; though many of them would be designated purely as "weathering phenomena" and studied as such by geologists.

Geomicrobiological processes are not very well documented, for which there are probably two reasons. The first is that, although fossil bacteria have been reported from rocks up to 2000 million years old[1], bacteria were not discovered until about 300 years ago; and the second is that many of these processes are extremely slow and thus not readily observable over a man's lifetime (though there is little doubt that the natural environment would be vastly different today but for the action of microbes on minerals over the past many millions of years).

Bacterial Iron and Sulphur Oxidation

The Organism

Oxidations of inorganic substances by microbes are described in some detail in Chapter 1 under *Nutrition;* like the much more common oxidation of simple organic substances such as sugars, their function is to yield reducing power for the microbe and energy for various synthetic reactions (such as carbon dioxide fixation in the organism we are about to consider), and for movement and other physiological processes. The substances involved in transferring reducing power and energy between various cell reactions are nucleotides (see also Chapter 1 for a brief account of their structure) such as nicotinamide-adenine dinucleotide (NAD) and adenosine-5′-triphosphate (ATP) respectively.

From an economic point of view the most important geomicrobiological process is the oxidation of metal sulphides. The chemolithotrophic bacteria which accomplish this are commonly regarded as belonging to the genus *Thiobacillus*. This genus, as its name suggests, is composed of organisms that have been shown to use the oxidation of sulphur or its more reduced compounds to obtain energy.

However, some thiobacilli that were also able to oxidize ferrous to ferric iron were isolated by Colmer and Hinkle[2] from acid waters in a coal mine in which pyrite (FeS_2) contained in the coal was being oxidized; they were subsequently designated *Thiobacillus ferro-oxidans*[3]. Various other strains resembling this original one have been isolated from copper-rich waters, streams, mines, waste dumps and soils in various parts of the world, where suitable conditions exist. It has since been reported that iron-oxidizing bacilli do not oxidize sulphur or its compounds, and thus a change of name to *Ferrobacillus ferro-oxidans* was proposed for the iron oxidizers. Until such conflicting evidence has been completely resolved, a final pronouncement cannot perhaps be made, though recent work[4] strongly indicates that various laboratory strains previously assigned either to *T. ferro-oxidans* or *F. ferro-oxidans* are identical, and different from *T. thio-oxidans*, *T. thioparus* and *T. denitrificans* (see Chapter 4, under *Combined Action of the Sulphur Bacteria in Metallic Corrosion*, for an account of the activities of these organisms) which together form a separate group. It seems reasonable to refer to our group as *Thiobacillus ferro-oxidans*, since this was its original name.

T. ferro-oxidans is a short, rod-like, acidophilic bacterium, about $1{\cdot}0 \times 0{\cdot}5\ \mu$, occurring singly or in pairs, and occasionally motile although flagella have not been seen under the microscope. It is aerobic, and does not form spores. Its optimum temperature for growth varies according to the substrate on which it is growing, and will also vary from one strain to another, but on the whole its active temperature range appears to be somewhat higher than that of many soil bacteria: growth will often take place at 45°C and is vigorous at 35°C. Commercially, most of the interest in this organism centres around its performance under field conditions, and the temperature at which a laboratory-grown isolate shows its maximum activity for a particular oxidation under well-controlled experimental conditions may not be ideal for a naturally-occurring strain under the almost unknown *in situ* conditions. The picture is made more complex by the natural process of adaptation to environment, which microbes readily undergo. It can be shown that if the temperature is gradually increased during a bacterial leaching experiment, the metal extraction rate may start to drop and then increase again owing to the effect of rise of temperature on the chemical leaching rate. Thus data are difficult to interpret; though the study of the temperature relations of leaching is important since the interiors of sulphide-containing dumps in which leaching is taking place are at elevated temperatures.

Mechanism of the Action

If we take pyrite as an example of a sulphide mineral, the following equations show a possible mechanism for oxidation:

$$2FeS_2 + 7O_2 + 2H_2O \rightarrow 2FeSO_4 + 2H_2SO_4 \qquad (1)$$

$$4FeSO_4 + O_2 + 2H_2SO_4 \rightarrow 2Fe_2(SO_4)_3 + 2H_2O \qquad (2)$$

$$FeS_2 + Fe_2(SO_4)_3 \rightarrow 3FeSO_4 + 2S \qquad (3)$$

$$2S + 3O_2 + 2H_2O \rightarrow 2H_2SO_4 \qquad (4)$$

The first reaction will take place in the absence of bacteria but the second one would be very slow in their absence since ferrous iron is reasonably stable in acid solution. The reaction shown in equation (3) will also take place in the absence of bacteria but, once again, the last reaction is speeded up at room temperature and normal pressure by the presence of sulphur-oxidizing bacteria. If another metal sulphide, for example covellite (CuS), is substituted for pyrite then copper sulphate would be obtained as an oxidation product as shown by equation (5):

$$CuS + Fe_2(SO_4)_3 \rightarrow CuSO_4 + 2FeSO_4 + S \tag{5}$$

This oxidation I define as an indirect oxidation. The bacteria are not "attacking" the mineral but are providing the chemicals (acid ferric sulphate – see equations (1) – (4)) which oxidize the metal sulphide. As acid ferric sulphate oxidizes different metal sulphides at different rates, the contribution of this part of the mechanism can be variable.

In addition to this indirect attack there is some indication that the bacteria will oxidize the sulphide directly. Ehrlich[5], using a synthetic chalcocite (Cu_2S), showed that bacteria had a strong influence on the rate of copper leaching in what he suggests is the absence of iron, although no analysis is given. Beck and Brown[6], using a different approach – a Warburg manometric technique – also obtained evidence indicating a direct as well as an indirect attack. A useful account of the solubilization of copper and iron ores has been given by Beck[7].

Some Examples of Mineral Attack

Work on a laboratory scale has shown that several metal sulphides can be directly or indirectly attacked by microbial processes. On such a scale the bacterial oxidation of metal sulphides can be investigated for example in shake flasks and columns. A small quantity of finely ground copper sulphide ore (about 5 g) is placed in a 250-ml cotton wool stoppered conical flask with a solution of certain inorganic salts (liquid medium). The composition of this medium is given in Table 7.1.

Table 7.1. *Inorganic salts medium for iron- and sulphur-oxidizing bacteria* (see Silverman and Lundgren[8])

$(NH_4)_2SO_4$	3·0 g
KCl	0·1 g
K_2HPO_4	0·5 g
$MgSO_4.7H_2O$	0·5 g
$Ca(NO_3)_2$	0·01 g
Distilled water to	700 ml
10*N* H_2SO_4	1·0 ml
$FeSO_4.7H_2O$	300 ml of 14·74% (w/v) solution

The solution is inoculated with suitable bacteria which can be obtained as freeze-dried cultures in glass ampoules (see Fig. 7.1) from the National Collection of Industrial Bacteria, the address of which is given in the *Appendix*, or can be isolated from an appropriate natural environment using an enrichment technique (the culturing of a sample of natural material in a medium designed to increase the population of the type of microbe it is desired to isolate relative to the total microbial population). The flask can then be placed on a shaker (Fig. 7.2) which keeps the bacteria in constant contact with fresh nutrients and ensures that the medium remains well supplied with oxygen (for aerobic respiration) and carbon dioxide (the sole carbon source) from the atmosphere.

This procedure for examining the attack on a mineral by a microbial process is rather unnatural: the bacteria enjoy well-chosen conditions of controlled temperature, and an adequate and constantly available supply of nutrients. To simulate a natural environment more effectively, columns containing larger particles of sulphide ore can be used (see Fig. 7.3). In these, as in nature, the above conditions are not necessarily present.

In passing, it could be mentioned that sulphur itself is a valuable mineral, and processes are being studied whereby sulphate-reducing bacteria (see Chapter 3), using a cheap mineral such as gypsum as a source of sulphate under anaerobic conditions, produce hydrogen sulphide on an industrial scale which could then be chemically oxidized to sulphur.

Uranium can be recovered from low-grade uranium ores, *e.g.*, brannerite ($(U,Ca,Fe,Th,Y)_3Ti_5O_{16}$), with the assistance of iron- and sulphur-oxidizing bacteria. However, there is no direct attack by the bacteria here. Ferric iron, from the pyrite which may be associated with the ore, oxidizes U^{IV} to U^{VI} which can then be removed in dilute acid solution (equation (6)).

$$UO_2 + Fe_2(SO_4)_3 \rightarrow UO_2SO_4 + 2FeSO_4 \quad (6)$$

The bacteria only oxidize the iron, providing the oxidant (ferric sulphate – see equation (2)). The acid ferric sulphate that is formed leaches the economic materials in the ore. The uranium can be recovered from these solutions by conventional ion exchange or solvent extraction techniques.

One question which is often raised when the commercial exploitation of these bacteria is considered concerns the possible toxic effect of the metals. Although various metals are necessary in biological systems as trace elements, when the concentration exceeds certain low limits the metal may be toxic or at least inhibitory. For example, copper and zinc in solution are known to be toxic to certain fish at 1 or 2 p.p.m. However, *Thiobacillus ferro-oxidans* is a robust organism. Although initially it is inhibited (but not killed) by comparatively low concentrations of, for example, copper (about 50 p.p.m.), it can be adapted over a period of time to tolerate much higher concentrations, of the order of 30 000 p.p.m. While rather similar figures can be presented for

Fig. 7.1 *Sealed glass ampoule of freeze-dried bacteria*

Fig. 7.2 *A constant-temperature shaker containing flask cultures of* Thiobacillus ferro-oxidans. (*Crown copyright reserved*)

Fig. 7.3 *Experimental leaching columns of various capacities, containing particles of low-grade uranium ore.*

An input of acidified water, or mineral salts solution, is provided, and partial recycling by means of a pump or air-lift. Effects of different pH and composition of mineral medium on leaching may be studied. (*Crown copyright reserved*) other metals such as uranium and zinc, certain others again are known to be highly inhibitory or toxic. Mercury is an example of a metal which is highly toxic and to which these bacteria seem unable to adapt.

There are other possible commercial uses for these iron and sulphur oxidizers if their activities can be economically harnessed: for example, recovery of metals other than copper and uranium from metal sulphides; the removal of sulphur from coal, shale or slag; and manganese recovery.

* * *

A different type of microbiological process which can result in mineral degradation is the leakage of metabolites, that is, intermediate substances formed during synthesis or respiration, from the microbial cell. Commercially, this type of process has only yet been exploited in a small number of fields such as the production of antibiotics, citric acid and other carboxylic acids, and a few extracellular enzymes, but it may in the future be extended to the extractive metallurgy field. For example Parés[9], using bacteria isolated from puddles on West African auriferous lateritic deposits, has shown that the gold can be brought into solution by the bacterial action. It appears that, given the correct nutrient medium, a metabolite is produced and excreted from the bacteria which will form a complex with the gold. Little is known about this metabolite except that it is a small-moleculed organic substance, soluble in butanol but not in chloroform, and contains no chlorine. This is then an example of an indirect attack by a bacterial metabolite.

An analogous situation occurs when ketogluconic acid is produced by, for example, *Pseudomonas* sp. Duff *et al.*[10] have shown that because of its chelation (a type of complex formation) with calcium, potassium, magnesium and other metals, the sugar acid could be a causative agent in "weathering" of natural silicates such as wollastonite ($CaSiO_3$) and chabazite ($CaAl_2Si_4O_{12} \cdot 6H_2O$), and in the solubilization of mineral phosphates in the soil.

Other workers have suggested different acids, such as citric, oxalic, lactic and phosphoglyceric acids, which could be biologically produced and could attack minerals and extract metals.

From an economic standpoint most interest has been centred on the bacterial leaching of uranium and copper; but it would be erroneous to suppose that bacteria are the only microbes that can be associated with mineral breakdown. Yeasts and fungi can also cause the release of metal ions from relatively insoluble minerals. Wenberg *et al.*[11] have reported the laboratory scale leaching of copper from copper-bearing minerals by fungi (*Penicillium*) under neutral or slightly alkaline conditions. Here too a chelating mechanism is suggested. This process is in marked contrast to the large scale biological leaching of sulphide copper using the acidophilic iron- and sulphur-oxidizing bacteria, where the pH is less than 3.

A recent Russian report mentions 30% of rock gold "turned into solution" while Ottow *et al.*[12], using various fungi including *Fusarium*, have shown iron reduction in α-hematite (Fe_2O_3). The phenomenon of removal of free iron oxides from soils and colloidal clays by biological reduction was reported as long ago as 1942 by Allison and Scarseth[13] but they did not isolate any microbes from the materials that they used for their experiments. Incidentally, these authors refer to a paper written in 1836 identifying microbes with the dissolution of iron compounds in nature.

* * *

In this contribution I am not going to comment on coal or petroleum which are allied to the metallurgical industry, although they both have a large number of microbial processes associated with their degradation.

In conclusion, I hope it has been shown that there is a future for these microbial processes. There are two facets which need examining very closely. Firstly there is the problem of the time factor, as was mentioned briefly at the beginning of this chapter. Man cannot wait millions of years to extract metals by microbiological techniques: these processes must therefore be optimized. That statement alone covers a frightening amount of work. Secondly, cheaper methods must be developed for recovering valuable products, in this case metals, from dilute solutions. It might be economical to recover copper from large volumes of solutions containing 100 p.p.m. of copper; it might be economical to recover uranium from solutions containing up to 15 p.p.m. of the metal; but what do we do with solutions containing 100 p.p.m. of zinc?

References

[1]BARGHOORN, E. S. AND TYLER, S. A. '*Micro-organisms from the Gun Flint Chert*'. Science, **147**, 563 (1965).

[2]COLMER, A. R. AND HINKLE, M. E. '*The role of micro-organisms in acid mine drainage: a preliminary report*'. Science, **106**, 253-256 (1947).

[3]TEMPLE, K. L. AND COLMER, A. R. '*The autotrophic oxidation of iron by a new bacterium,* Thiobacillus ferro-oxidans'. J. Bact., **62**, 605-611 (1951).

[4]HUTCHINSON, M., JOHNSTONE, K. I. AND WHITE, D. '*Taxonomy of the acidophilic thiobacilli*'. J. Gen. Microbiol., **44**, 373-381 (1966).

[5]EHRLICH, H. L. '*Biogeochemistry of sulfur isotopes*'. In National Science Foundation Symposium, ed. by H. L. Jensen, Yale University 1962, 153–168.

[6]BECK, J. V. AND BROWN, D. G. '*Direct sulfide oxidation in the solubilization of sulfide ores by* Thiobacillus ferro-oxidans'. J. Bact., **96**, 1433–1434 (1968).

[7]BECK, J. V. 'Thiobacillus ferro-oxidans *and its relation to the solubilization of ores of copper and iron*'. In Fermentation Advances, ed. by D. Perlman, 747–771. London and New York: Academic Press, 1970.

[8]SILVERMAN, M. P. AND LUNDGREN, D. G. '*Studies on the chemo-autotrophic iron bacterium* Ferrobacillus ferro-oxidans'. J. Bact., 77, 642–647 (1959).

[9]PARÉS, Y. '*Solvent extraction of gold by bacteria*'. Revue de l'Industrie Minérale, **50**, 408–415 (1968).
[10]DUFF, R. B., WEBLEY, D. M. AND SCOTT, R. O. '*Solubilization of minerals and related materials by 2-ketogluconic acid-producing bacteria*'. Soil Science, **95**, 105–114 (1963).
[11]WENBERG, G. M., ERBISCH, F. H. AND VIOLIN, M. E. '*Leaching of copper by fungi*'. American Institute of Mining and Metallurgical Engineering Meeting, February 1969, Washington, D.C. (Preprint 69B 43).
[12]OTTOW, J. C. G. AND VON KLOPOTEK, A. '*Enzymatic reduction of iron oxide by fungi*'. Appl. Microbiol., **18**, 41–43 (1969).
[13]ALLISON, L. E. AND SCARSETH, G. D. '*A biological reduction method for removing free iron oxides from soils and colloidal clays*'. J. Amer. Soc. Agronomy, **34**, 616–623 (1942).

CHAPTER EIGHT: A. W. FLETCHER

Copper Recovery from Low-Grade Ore by Bacterial Leaching

Introduction

As we have already seen (Chapter 1), microbes are ubiquitous on this planet, and the layman may assume that all are harmful and cause disease; yet the truth is that most micro-organisms are harmless and many are necessary for the survival of other living things. The diseases of man and the microbial metabolism of organic matter received early emphasis by microbiologists, and man's dependence on agriculture gave impetus to soil microbiology which has led to a better understanding of the important role played by bacteria in oxidizing and reducing minerals to soluble forms that can be utilized by plants. In the early years of the development of the subject, the inorganic environment was virtually ignored, although in more recent times geologists have become aware of the important role played by microbes in the formation of many sedimentary deposits. The idea that a bacterial process can assist in the leaching of metals from ores for economic recovery of the metal is a comparatively recent one, dating from the mid-1950's. Long before *Thiobacillus ferro-oxidans* was identified, dumps of mine waste containing low-grade copper ore were being watered, gradually to solubilize the copper by what was then considered to be a natural leaching process.

It is of interest to consider why the subject has received so much attention in recent years. This has certainly been due to the explosion in demand for metals by the industrial society during this century. It is estimated that world mineral consumption over the last decade equals the total amount consumed throughout all man's previous existence as a user of minerals. Based on a U.S.A. Government estimate, demand for minerals and metals is likely to double in the next ten years and treble in the next twenty years. By now the accessible high-grade deposits of many minerals have been thoroughly exploited and it is necessary to consider lower and lower grade and complex deposits as ore resources for the future. This downward trend will continue and the availability and stability of the price of minerals and metals will depend largely on further advances in the technology of extraction. It is within this context that we should consider the use of dump leaching, in which a bacterial process plays an important role and which is already being used for the extraction of some 227 000 tonnes of copper per annum in the United States.

During the last century copper ore could only be mined and processed economically if it contained 10% or more of the metal. With the turn of the century came improvements in mineral separation techniques and the discovery of flotation, still the most widely used method of mineral processing. By the 1920's this had been developed to an extent which permitted economic recoveries from material containing as little as 2% copper. Within the last decade further advances in the technology of mining and mineral processing, including flotation, have lowered the economic grade to some 1% copper. With bacterial leaching, however, copper can be economically recovered from mine waste which contains as little as 0·25 to 0·5% copper.

The Activities of *Thiobacillus ferro-oxidans*

The bacteria involved in these leaching reactions—*Thiobacillus ferro-oxidans*—have the ability to use the oxidation of insoluble metal sulphides to water-soluble metal sulphates, and the oxidation of ferrous iron in solution, as sources of energy for growth (that is, they are chemolithotrophs). They require oxygen, use carbon dioxide but not organic matter as a source of carbon for cell growth, and they can also use ammonium nitrogen. Generally they are regarded as being nutritionally undemanding. These bacteria are only about 1 micron in length and can, therefore, only be seen under a high-powered microscope. They were first identified in acid mine waters in the U.S.A. in 1954 but since that time they have been found to be ubiquitous, having been identified in many parts of the world including the U.K. The chemical and biological reactions involved in the dissolution of metal sulphides are complex and are not completely understood but it is generally agreed that direct and indirect processes are involved.

In the direct oxidation of copper sulphides the bacteria must be in physical contact with the mineral in order to catalyse the oxidation reaction:

$$\underset{\text{(chalcopyrite)}}{CuFeS_2} + 4O_2 \rightarrow CuSO_4 + FeSO_4$$

In the indirect process the bacteria oxidize ferrous iron in solution to ferric iron which then chemically attacks the sulphide mineral:

$$CuFeS_2 + 2Fe_2(SO_4)_3 + 2H_2O + 3O_2 \rightarrow CuSO_4 + 5FeSO_4 + 2H_2SO_4$$

Malouf and Prater[1] have described the results of investigations initiated by the Kennecott Copper Corporation on the use of these bacteria for the leaching of copper from low-grade ores.

It was found that a very effective lixiviant for copper sulphide minerals can be generated by using bacterial oxidation to oxidize pyrite to sulphuric acid and ferrous sulphate, and the same bacterial oxidation to convert the ferrous iron in solution to ferric iron. The acid obtained in the oxidation of the pyrite is sufficient to hold the ferric sulphate in solution until it comes into contact with the copper sulphide minerals. These minerals are oxidized to produce copper sulphate, and the ferric sulphate is reduced to ferrous sulphate to provide again the energy source for bacterial oxidation.

The heat generated by the oxidation of the minerals has an accelerating effect on the leaching of copper from mineral waste. It has been noted that in one large waste dump where the leaching of copper had been started with ferric sulphate solution, sufficient heat was generated to raise the temperature of the water by 25 °F (14 °C). Controlled laboratory tests have indicated that this increase in temperature increased the leaching rate threefold over that obtained with systems at ambient temperatures.

The bacteria found in mine waters exist under a fairly wide range of environmental conditions but, in general, the mine water is weakly acidic and contains low concentrations of various salts. However, in the laboratory it has been possible to increase the tolerance of bacteria of this species to high acid and salt concentrations in solutions used for leaching. Similarly, it has proved possible to increase the rate of bacterial oxidation markedly above that observed in nature. Active bacterial cultures have been established in solutions containing as much as 15 gram/litre of copper, as much as 40 g/l of iron, and in sulphuric acid concentrations sufficiently high to yield an indicated pH of 0·5. Similarly, in the laboratory, iron oxidation rates have been obtained as high as 20 g iron converted to the ferric state per litre of solution per 24 h and pyrite dissolution rates as high as 1·2% of the pyrite present in the system per 24 h. Under these conditions the bacterial activity is far greater than that encountered in nature or in processes employing corresponding chemical oxidation based solely upon air and water.

The oxidizing bacteria can exist under a wide range of concentration of ions in solution although they are sensitive to sudden and drastic changes in the composition of the solutions which constitute their environment. Tolerance to a relatively high concentration of ions which normally would sterilize a culture can be developed by gradually increasing the concentration of the ion in solution. A notable example of this was found with the zinc ion. It was observed that concentrations of zinc as low as 50 p.p.m. were adequate to kill the bacteria occurring in mine water, although tests showed that the oxidation of sphalerite and solubilization of the zinc by bacterial oxidation was possible. Tolerance for zinc ion was obtained by contacting a viable culture with sphalerite, thus effecting a slow increase in zinc ion concentration to which the bacteria became adapted. Similar adaptations to high metal ion concentrations have been observed in the case of copper, which is toxic to the bacteria unless a tolerance has been developed by adaptation. By slowly developing the tolerance for copper, it has been possible to produce active bacterial cultures in solutions containing as much as 15 g/l of copper.

Bacterial cultures become inactive, but are not destroyed, when exposed to direct sunlight. Experiments have been made to evaluate bacterial activity in open ponds with solution depths of up to 2 ft (0·61 m). Virtually no bacterial oxidation was noted in these ponds, even though the solutions were kept in darkness for 50% of the time and viable bacteria were present. It appears that in shallow ponds sufficient radiation is transmitted through up to 2 ft of water to inhibit bacterial activity, though it does not destroy the bacteria present. The bacteria are extremely sensitive to ultraviolet light, for only brief exposure to this radiation will completely sterilize a bacterial culture.

Temperature has been found to exert a very pronounced effect on the rate of bacterial conversion of Fe^{2+} to Fe^{3+} with maximum bacterial activity occurring at approximately 35 °C. Temperatures above 50 °C have been found to destroy the bacteria.

The oxidizing bacteria are active only in acid media. In general, bacterial action is most pronounced in media having a pH of between 2·0 and 3·5. Both above and below this pH range the rate of bacterial oxidation decreases, and at pH values above 6·0 bacterial action is almost completely inhibited. The bacteria are destroyed in alkaline media (pH 9).

Laboratory tests indicate that bacteria concentrate at the surface of solids where they build up large colonies. Crushed rock particles provide a surface for such colonies and, as a result, markedly affect oxidation rates. The supply of oxygen to the bacteria, particularly in large dumps, is in some respects the most important factor determining their activity. This can be supplied by the circulation of aerated solutions to the site of bacterial activity; but in large dumps, where percolation rates limit the volume of water which can be passed through the dump, the oxygen obtained from air passing through the dump by natural convection is probably equally important.

Application of Bacterial Leaching

By far the most important application of the process during the last 10 to 20 years has been the recovery of copper from low-grade ore in the south-western United States. As ore grades have fallen the tonnage of ore and waste mined has increased, and this has been made possible by the development of open pit mining and improvements in earth-moving equipment. For example, 182-tonne rear dump diesel-electric trucks are not uncommon. In 1900, mines in this region were no bigger than 455 t.p.d. (tonnes per day), but today the smallest is 3640 t.p.d., approximately eleven are about 10 900 t.p.d. and the largest are approaching 91 000 t.p.d. Easily the giant of all leaching operations is that at Kennecott's Bingham Canyon pit where mine waste dumps are said to contain over $3{\cdot}6 \times 10^9$ tonnes and are being added to at the rate of 245 000 t.p.d. Over 7·5 miles (12 km) of epoxy-lined asbestos-cement pipe handle a flow of 40 000 gal (151 m^3) per minute. In the U.S.A., copper production by dump leaching has increased from 136 000 tonnes in 1965 to 227 000 tonnes in 1968.

Current leaching practice is an approximation of natural processes in which old dumps of low grade ore, considered waste, are irrigated with water or spent leach liquor, sometimes acidified with sulphuric acid. The water percolates through the dump dissolving copper as it descends and the solution is passed over scrap iron in launders where copper powder is precipitated by the cementation reaction:—

$$Cu^{2+} + SO_4^{\;2-} + \text{Fe(metal)} \rightarrow \text{Cu(metal)} + Fe^{2+} + SO_4^{\;2-}$$

The mine dumps that are leached are enormous piles of uncrushed rock stripped from disseminated porphyry type copper deposits. The leach dumps average about 0.3% copper in the form of a variety of sulphide and oxide minerals. Chalcopyrite is the principal sulphide mineral in most dumps and chalcocite in a few dumps. Minor amounts of bornite, covellite and enargite may be present. Pyrite is a common accessory mineral. The oxidized copper minerals, present in the dumps to a greater or lesser degree, are chiefly malachite, azurite and chrysocolla.

Most dumps are deposited upon the existing topography and were not specially designed for percolation leaching. Where possible, the dumps are located on impermeable ground with a drainage pattern enabling easy collection of leach solution. When such conditions for dump emplacement are not available, nearby sites are specially prepared. One company, for example, covered an area more than a mile from the mine with an asphaltic pad four inches (10.2 cm) thick to prevent seepage losses of copper solutions into the alluvium. For convenience, large dumps are usually raised in lifts of 50–100 ft (15·25–30·5 m).

Leaching solutions are pumped to a variety of distribution systems for spreading over the surface of a dump. Plastic-lined asbestos-cement or stainless steel pipes are commonly used throughout circulation systems. Common distribution methods are flooding of the dump surface in a series of ponds, spraying of the solution from metal or plastic heads, and injection of the solution into the dumps through perforated plastic pipe.

Dumps may be leached in cycles in which solutions are allowed to percolate for periods of at least several months after which the dumps are allowed to stand without liquid addition for a period of aeration. Other types of operations maintain steady percolation over very long periods.

Perhaps the most significant improvement in the rate of copper extraction came from the discovery that the leaching reactions were accelerated by adding sulphuric acid and thus lowering the pH of liquor recycled from

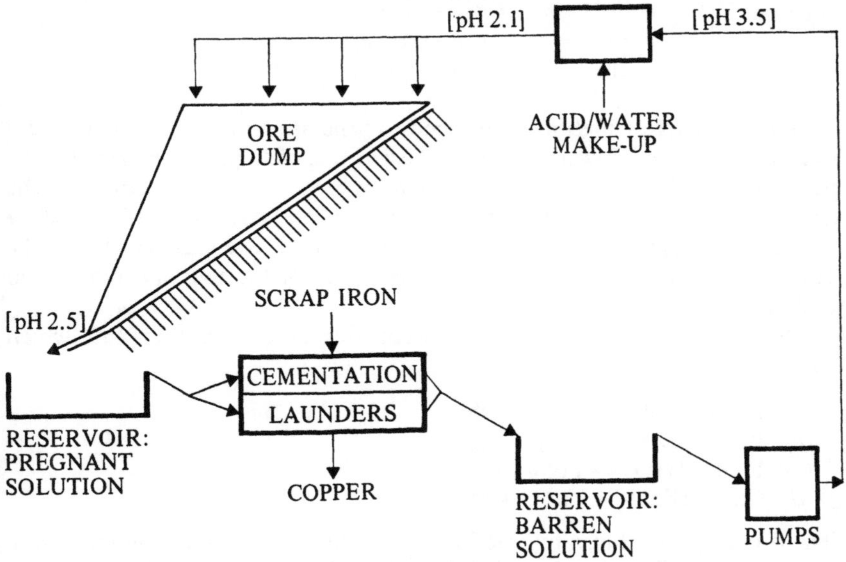

Fig. 8.1 *A typical flowsheet for copper recovery from a dump of low-grade ore by bacterial leaching*

cementation from 3·5 to between 1·9 and 2·4. This is now standard practice at most copper dump leaching operations in the south-western United States; Fig. 8.1 shows this diagrammatically. Important improvements in copper recovery have been made by recognizing that the temperature of the recycled liquor should be maintained at as high a level as is practicable, and that adequate access of air to the interior of the dump is essential for bacterial growth and the efficiency of leaching.

At Kennecott it is believed that ferric salts play an important role in the leaching reactions and it is now thought that vigorous oxidation of $Fe^{2+} \rightarrow Fe^{3+}$ takes place in the upper areas of the dump and that the addition of acid to the liquor prevents the deposition of basic ferric salts which would gradually reduce the permeability of the dump. The iron content of the leach liquor is of the order of 2–4 g/l and there is evidence to show that higher concentrations are undesirable. Although iron is continually added to the recycling liquor at the cementation stage, surprisingly there is no build-up of iron salts—no doubt because of the precipitation of hydrolysed iron compounds.

There is considerable variation in the composition of leach liquors at different operations, as pointed out in the survey carried out by Evans and Sheffer[2]. Typical analyses are given in Table 8.1.

Table 8.1. *Range of composition of leach liquors*

Copper	0·75 – 2·16 g/l
Total iron	0·3 – 6·6 g/l
Fe^{2+}	0·1 – 3·6 g/l
Fe^{3+}	0·2 – 3·0 g/l
pH	1·9 – 3·0 (mostly 2·5)

Dumps relatively high in pyrite tend to generate more sulphuric acid and iron salts than those lower in pyrite. The addition of sulphuric acid affects the pH of the leach solution and the amount of ferric iron salts retained in the liquor. Solutions relatively high in sulphuric acid and ferric sulphate are better solvents for copper minerals than those of lower strength. However, the economics of acid addition must be determined for the special conditions in each operation.

The important reactions which occur during cementation and which consume iron are:

(*a*) $Fe + CuSO_4 \rightarrow Cu + FeSO_4$
(*b*) $Fe + Fe_2(SO_4)_3 \rightarrow 3FeSO_4$
(*c*) $Fe + H_2SO_4 \rightarrow FeSO_4 + H_2$
(*d*) $Cu + 2Fe^{3+} \rightarrow Cu^{2+} \rightarrow 2Fe^{2+}$

The consumption of iron by reactions (*b*), (*c*) and (*d*) is wasteful and therefore these reactions should be impeded if at all possible. Reaction (*a*) proceeds at a faster rate than (*b*) and (*c*) and within the temperature range 15 °–34 °C the rate increases with rise of temperature. Furthermore the rate of reaction (*a*) is controlled by the diffusion of reactants to the iron surface and the diffusion of products away from the surface. These principles have been employed by the Kennecott Copper Corporation in the design of a Cone Precipitator for the cementation of copper from dump leaching solutions[3]. This design has effectively increased the rate of copper precipitation so that barren liquor can be returned to the leaching dumps at a higher temperature. In addition, copper recoveries from solution have been increased and iron and labour costs reduced.

As an alternative to cementation, solvent extraction is now in use at two mines to obtain a pure concentrated copper sulphate solution from dump leach liquor for the production of high grade electrolytic copper.

At the Butte plant in Montana the leaching of dumps specially built on a prepared base was inaugurated in 1964. The process is similar to that already described but deserves special mention because it is operated successfully in a northern latitude where climatic conditions are very different from the semi-desert conditions which prevail in the south-western United States. With temperatures ranging from severe cold in winter to very hot summers it is apparent that the climate is more severe than that of the U.K. An interesting innovation is the effort that is made to keep the dumps well aerated. The dumps are churn-drilled on a grid pattern, and into each hole is inserted a 4 inch (10·2 cm) diameter PVC pipe which is perforated with four ½ inch (12·7 mm) holes spaced 90° apart at 1 ft (0·3 m) intervals throughout its entire length. These pipes are spaced 50 ft (15·2 m) apart in parallel rows which are 25 ft (7·6 m) apart. Paradoxically, the metal content of liquor draining from the dumps is higher in winter than in summer. This is attributed to the fact that while the interior of the dump is at a constant temperature of 26 °C at all seasons, the higher temperature differential between the dump and the atmosphere in winter conditions results in an increased air flow through the dump. In sub-zero temperatures, air from each of the holes reaches a volume of 400 ft^3 (11.3 m^3) per minute and its oxygen content is as low as 4%.

At the Miami mine in Arizona a different leaching system is used to recover copper from low-grade ore left behind after deep mining[4]. The method of mining used for the recovery of rich ore for processing by conventional methods was block caving, and this left behind some copper in stopes, crushed pillars etc. In addition there was a capping of oxide copper which was not mined. The old underground mine workings have been skilfully used as drainage channels and ponds for the collection of liquor percolating through the low-grade ore; this is shown diagrammatically in Fig. 8.2. At this mine the surface to be leached had an area of 5×10^6 ft^2 (0·46 km^2) and in twenty years half the area was treated. From the time watering sprays are turned on at the surface, the liquor takes 3–4 weeks to penetrate 600 ft (183 m) of ore and start

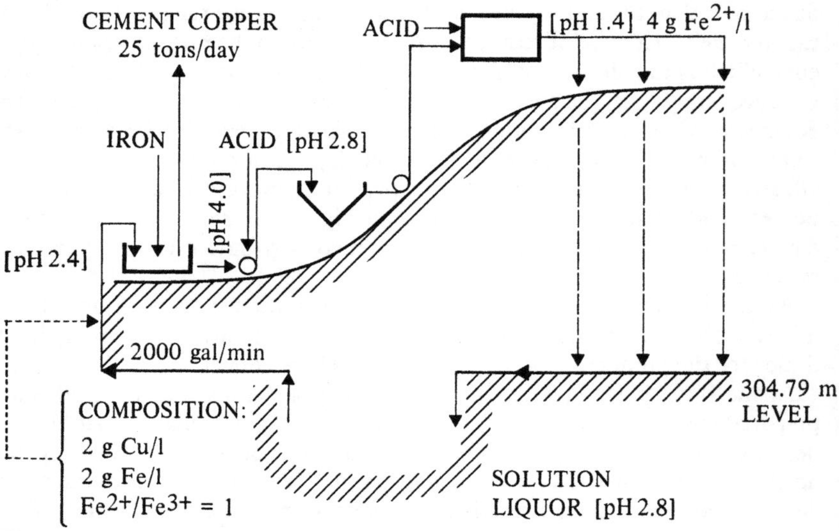

Fig. 8.2 In situ *copper leaching at the Miami mine, Arizona*

collecting in mine workings at the 1,000 ft (305 m) level. It is found that copper extraction from the ore is increased by periodic watering with an interval of about three months between watering campaigns. Iron and acid consumption at the mine is 1·3 and 2·4 lb per lb (or kg per kg) of copper respectively.

Allman and Harris[5] state that the watering rate at Rum Jungle was 0·11 gal/ft^3 h (17·6 l/m^3 h), while at the Anaconda Company's Butte Plant in Montana the rate is stated to be 0·13 gal/ft^2 h (5·3 l/m^2 h).

Economics

Heap leaching is probably the only metal extraction process in which the time involved from mining the ore to metal production is measured in months rather than hours or at the most days. Metal recovery cannot be accurately forecast and will never approach that of more conventional milling processes. Against these disadvantages are the low capital and operating costs. Because of differences in the mineralogical composition and size of the ore and differences in the size and shape of dumps, watering procedures etc., it is not surprising that there are great variations in reported leaching rates. For example, the copper extraction rate can vary from 3–10% per annum, giving 50% copper recovery in approximately 16 and 5 years respectively.

Cooper[6] has provided data (Table 8.2) foı the costs of *in situ* leaching in Arizona for the production of 9000 short tons (8180 tonnes) of copper

Table 8.2. *Costs of* in situ *leaching and mining of copper ore in Arizona*

	Cents per lb	Cents per kg
In situ leaching:		
Labour	0·93	2·05
Acid	3·25	7·17
Iron	3·36	7·41
Pumps and piping	0·08	0·18
Electricity	0·28	0·63
Miscellaneous: amortization, freight, maintenance, overheads	2·10	4·63
	10·00	22·07
Mining:	6–12	13–26

Note: 1 U.S.A. gal/min (3·78 l/min) of solution containing 1 g/l Cu flowing continuously for 1 year will supply 4380 lb (1986·7 kg) Cu or nearly 2 tonnes, which at £500/tonne is worth £1000.

equivalent per annum. The copper from dump leach liquors is precipitated on iron to produce cement copper containing 70–80% Cu.

A more detailed assessment of the economics of heap leaching has been made by Shoemaker and Darrah[7], but the calculations are based on acid leaching of oxide ore.

The successful results obtained from the dump leaching of low-grade copper ores have inspired mining and metallurgical authorities to consider the wider possibility of mining metals from ores by the direct leaching of deposits as they occur underground in nature. Thus terms such as "in place" or "*in situ*" leaching and "solution mining" are now commonplace. It has been proposed that the problem of creating zones of permeability adequate to permit intimate contact between leach solutions and ore minerals can be overcome by the use of nuclear explosives deep in the ore body. Detailed plans (Project Sloop) have been made jointly by the Kennecott Copper Corporation and the U.S. Atomic Energy Commission to apply the technique to the recovery of copper from a deposit near Safford, Arizona[8]. It is estimated that 1520-kiloton contained nuclear devices will be required to fracture the ore body and make it suitable for *in situ* leaching. A recent analysis[9] of process economics indicates that the technique could be commercially viable provided that the technical difficulties associated with the use of nuclear explosives can be overcome. If the experiment succeeds there are exciting prospects ahead for the use of bacterial leaching in the winning of metals from ores.

References

[1]MALOUF, E. E. AND PRATER, J. D. '*Role of bacteria in the alteration of sulphide minerals*'. Journal of Metals, **13**, 353–356 (1961).

[2]EVANS, L. G. AND SHEFFER, H. W. '*Dump and heap leaching practices in the Western United States*'. Mining Congress Journal, **54** (9), 96–100 (1968).

[3]SPEDDEN, H. R. *et al.* '*Use of cone-type precipitators to recover copper from copper-bearing solutions*'. American Institute of Mining and Metallurgical Engineering Annual Meeting, March 1966 (Preprint 66B 86).

[4]FLETCHER, J. B. '*In place leaching in Miami mine*'. American Institute of Mining and Metallurgical Engineering Arizona Section Meeting, April 1962.

[5]ALLMAN, M. B. AND HARRIS, J. A. '*An experiment in heap leaching*'. Mining Congress Journal, **55** (7), 28–31 (1969).

[6]COOPER, F. D. '*Copper hydrometallurgy—a review and outlook*'. U.S. Bureau of Mines Information Circular 8394 (1968).

[7]SHOEMAKER, R. S. AND DARRAH, R. M. '*The economics of heap leaching*'. Mining Engineering, **90,** 68–70 (1968).

[8]ANON. '*Seek profitability answer to nuclear* in-situ *copper leaching at Safford*'. World Mining, **21** (1), 48–51 (1968).

[9]ANON. '*Buried nuclear fracturing continues to be the most logical mining use of nuclear explosives*'. World Mining, **23** (3), 50–54 (1970).

Appendix

Classification of the microbial genera mentioned in the text

Bacteria
Fungi (moulds and yeasts)
Algae
Protozoa
Culture Collections

Bacteria

Family	Genera
Caulobacteraceae	*Gallionella*
Pseudomonadaceae	*Pseudomonas*
Thiobacteriaceae	*Thiobacillus*
	(*Ferrobacillus*)
Spirillaceae	*Vibrio*
	Desulfovibrio
(Uncertain status)	*Desulfotomaculum*
Achromobacteraceae	*Achromobacter*
	Alcaligenes
	Flavobacterium
Bacillaceae	*Bacillus*
	Clostridium (NOT "*Clostridium nigrificans*")
Brucellaceae	*Haemophilus*
Enterobacteriaceae	*Aerobacter*
	Proteus
Micrococcaceae	*Micrococcus*
Chlamydobacteriaceae	*Sphaerotilus*
	(*Clonothrix*)
	(*Crenothrix*)
	(*Leptothrix*)
Mycobacteriaceae	*Mycobacterium*
Actinomycetaceae	*Nocardia*

Fungi (moulds and yeasts)

Class	Genera
Ascomycetes	*Aspergillus*
Fungi Imperfecti	*Alternaria*
	Cladosporium
	Fusarium
	Monilia

Fungi Imperfecti cont.	*Paecilomyces*
	Penicillium
	Trichoderma
Yeasts	*Candida*
	(*Torula*)
	Rhodotorula

Algae

Division	Genera
Cyanophyta	*Chroococcus*
	Oscillatoria
Bacillariophyta	*Navicula*
Euglenophyta	*Euglena*
Chlorophyta	*Chlorococcum*
	Scenedesmus
	Spirogyra
	Ulothrix
Charophyta	*Chara*

Protozoa
Genus: *Amoeba*

* * *

Culture collections

Cultures of micro-organisms may be purchased from various sources in the UK, including the following culture collections.

Bacteria: The National Collection of Industrial Bacteria,
Torry Research Station,
135 Abbey Road,
Aberdeen,
Scotland

Moulds: The Commonwealth Mycological Institute,
Ferry Lane,
Kew,
Surrey

Yeasts: The National Collection of Yeast Cultures,
Brewing Industry Research Foundation,
Nutfield,
Surrey

Algae: The Culture Centre of Algae and Protozoa,
36 Storey's Way,
Cambridge

Index

GPSR Compliance

The European Union's (EU) General Product Safety Regulation (GPSR) is a set of rules that requires consumer products to be safe and our obligations to ensure this.

If you have any concerns about our products, you can contact us on ProductSafety@springernature.com

In case Publisher is established outside the EU, the EU authorized representative is:

Springer Nature Customer Service Center GmbH
Europaplatz 3
69115 Heidelberg, Germany

Batch number: 10584446

Printed by Printforce, the Netherlands